AN INTRODUCTION TO
ENGINEERING MATHEMATICS

AN INTRODUCTION TO
ENGINEERING
MATHEMATICS

by

D. McMULLIN, B.A.,
Sometime Examiner to the Surrey Higher Education Committee

and

A. C. PARKINSON, A.C.P. (Hons.), F.Coll.H.,
Author of "Intermediate Engineering Drawing", "Engineering
Workshop Drawing", "First Year Engineering Drawing"

Lecturers in the Technical College,
Kingston-upon-Thames

CAMBRIDGE
AT THE UNIVERSITY PRESS
1936

CAMBRIDGE
UNIVERSITY PRESS

University Printing House, Cambridge CB2 8BS, United Kingdom

Cambridge University Press is part of the University of Cambridge.

It furthers the University's mission by disseminating knowledge in the pursuit of
education, learning and research at the highest international levels of excellence.

www.cambridge.org
Information on this title: www.cambridge.org/9781316611906

© Cambridge University Press 1936

First published 1936
First paperback edition 2016

A catalogue record for this publication is available from the British Library

ISBN 978-1-316-61190-6 Paperback

PREFACE

This book is intended primarily for the use of students of Practical Mathematics in so far as it comprises a subject taken in the first year of the senior technical courses, established in Technical Colleges and Institutes.

The syllabuses prescribed in different areas are not precisely alike in scope and standard. We have endeavoured, therefore, with the kind assistance of several experienced colleagues, in different parts of the country, to establish an L.C.M. of general requirement as to syllabus content. It is hoped that, in relation to space available, the range of subject matter will prove adequate to the main purpose of the book. As an example, mention may be made of the inclusion of quadratic equations, which appear in some first-year examination papers.

This book has a reasonably practical bias. Technical students having daily contact with things of practical significance are apt to become impatient of purely academic study—desiring above all to relate principles to practice, to express the academic in terms of direct utility. At the same time, however, their practical work is usually so briefly specialised that, in our view, technical allusions and practical problems beyond their range of experience tend to defeat their own object. With this in mind we have endeavoured to draw from a long aggregate experience in engineering and teaching, and to give expression to this experience, in a sanely balanced course in the groundwork of mathematics, which will enable a student to retain easy contact with the pure mathematics of his pre-technical studies and yet find a sufficiency of generally understood technical allusions, illustrations, and graded applications to awaken interest, and enliven endeavour.

The leading geometrical propositions commonly included in the syllabuses have been mentioned, together with a sufficiency of practical verifications or brief deductive proofs. Modern tendency appears to favour an extension of the time spent in deductive geometry, the value of which, in comparison with the slipshod empiricism which often arises from undue dependence upon graphical verifications, cannot be overlooked. Nevertheless, our experience is that, so far as evening classes are concerned, the available time is insufficient for much greater elaboration of this branch of the work. We have aimed at logical sequence of leading theorems, clear deductions and concise, practical applications.

The ground covered also includes the work taken in Junior Technical Schools in Practical Mathematics during the first and second years.

Thanks are tendered to the following for ready and courteous permission to reprint questions set in examinations: Northern Counties Technical Examinations Council; Union of Educational Institutions; City and Guilds of London Institute; Royal Society of Arts; Surrey Education Committee; H.M. Stationery Office.

D. McMULLIN

A. C. PARKINSON

April 1936

CONTENTS

Chapter I. Factors and Multiples; Vulgar Fractions; Regular Solids; Mensuration of Rectangle and Triangle; Decimals; Approximations and Significant Figures; Revision Exercises   pp. 1–29

Chapter II. Ratio and Proportion; Inverse Proportion; Averages; Percentages   pp. 30–38

Chapter III. Powers and Roots   pp. 39–41

Chapter IV. Symbols; Algebraic Sums; Algebraic Multiplication; Algebraic Division; Rule of Signs; Brackets and Symbolic Expression   pp. 42–51

Chapter V. Factors   pp. 52–56

Chapter VI. Simple Equations; Problems; Change of Subject   pp. 57–64

Chapter VII. Angles and Angle Measure; Parallelism; Triangles; Angles of Polygons; Congruence of Triangles   pp. 65–81

Chapter VIII. Geometry and Mensuration of Quadrilaterals and Polygons; Theorem of Pythagoras   pp. 82–99

Chapter IX. Logarithms   pp. 100–9

Chapter X. Mensuration of the Circle, Sector, Annulus   pp. 110–18

Chapter XI. Use of Squared Paper; Proportion by Graphs   pp. 119–26

Chapter XII. Simultaneous Equations; Problems   pp. 127–32

Chapter XIII. Co-ordinates, Straight Lines and Slopes; Graphical Solution of Simultaneous Equations; The Equation of a Straight Line   pp. 133–46

Chapter XIV. Mensuration of Solids; Volumes and Surface Areas of Cubes and Prisms; Cylinders, Pyramids, Cones, Frusta; Spheres; Similar Figures, Plane and Solid   pp. 147–75

Chapter XV. Curved Graphs   pp. 176–78

Chapter XVI. Quadratic Equations   pp. 179–84

CONTENTS

Chapter XVII. Trigonometry—Ratios and Exercises   pp. 185–95

Chapter XVIII. Geometry of the Circle; Chords, Angles and Tangency; Useful Propositions and Exercises   pp. 196–211

Chapter XIX. Loci   pp. 212–15

Tests   pp. 216–30

Tables and Constants   pp. 231–32

Mensuration Formulae (Summary of)   pp. 233–37

Tables. Logarithms; Antilogarithms; Sines; Cosines; Tangents   pp. 238–47

Index   p. 248

Answers   pp. 249–66

CHAPTER I

FACTORS

1. A **factor** *is a number which will divide into another number an exact number of times.* Thus 3 is a factor of 15.

If the same factor will divide into two or more numbers, it is said to be a **common factor** of these numbers. Thus 4 is a common factor of 16 and 24.

Again considering these two numbers 16 and 24, we see that 8 is also a common factor. It is also the greatest factor of both numbers, it is therefore called the **highest common factor**, written H.C.F.

A **Prime Number** *is one whose only factors are itself and unity.* Thus 11, 13, 19, 23 are prime numbers. A **prime factor** is one which in itself is a prime number.

To find the H.C.F. of two or more numbers

(1) Find the prime factors of each number, (2) put down all the factors common to each number. The product of these common factors is the H.C.F. required.

EXAMPLE. Find the H.C.F. of 105, 180, 210.

$$105 = 3 . 5 . 7,$$
$$180 = 2 . 2 . 3 . 3 . 5,$$
$$210 = 2 . 3 . 5 . 7.$$

The factors that are common to these prime factors are 3 and 5. Therefore the H.C.F. is 15.

Exercises 1

Find the H.C.F. of:

1. 63, 147.
2. 98, 70.
3. 51, 57.
4. 12, 15, 30, 72.
5. 60, 105, 195, 210.
6. 140, 780, 550.
7. 120, 360, 480.
8. 210, 1155.
9. 462, 112.
10. 600, 336.
11. 420, 150, 630.
12. 171, 152, 133.

13. There are three rods, 3 ft. 4 in., 8 ft. 4 in. and 8 ft. 9 in. long respectively. What is the greatest common length that may be cut off these rods an exact number of times? How many of these lengths will there be altogether?

14. A courtyard 30 ft. long and 25 ft. 6 in. broad is to be paved with square tiles. What will be the dimensions of the tiles so that the least number may be used?

MULTIPLES

2. *A number which contains another number an exact number of times is said to be a* **multiple** *of that number*. Thus 20 is a multiple of 2.

When a number contains two or more other numbers an exact number of times, it is said to be a **common multiple** of these numbers. Thus 20 is a common multiple of 2 and 5.

Again 10 is a common multiple of 2 and 5, and it is also the smallest or least number or multiple which exactly contains 2 and 5. It is therefore called the **least common multiple** of 2 and 5. **L.C.M.** denotes least common multiple.

To find the L.C.M. of two or more numbers

1. Write down prime factors of all the numbers.
2. Write down the prime factors of the largest number.
3. With these factors, place any factors of the other numbers not already there.
4. The product of the factors so obtained is the L.C.M. required.

EXAMPLE 1. Find the L.C.M. of 4, 12, 15.

(1)
$$15 = 3 \cdot 5.$$
$$12 = 2 \cdot 2 \cdot 3.$$
$$4 = 2 \cdot 2.$$

(2) Prime factors of largest number:
$$3 \cdot 5.$$

(3) With factors in (2) place the factors of 12 and 4 not already there:
$$3 \cdot 5 \cdot 2 \cdot 2 = 60.$$

∴ the L.C.M. of 4, 12, 15 is 60.

EXAMPLE 2. Find the L.C.M. of 8, 12, 15, 16:
$$8 = 2 \cdot 2 \cdot 2.$$
$$12 = 2 \cdot 2 \cdot 3.$$
$$15 = 3 \cdot 5.$$
$$16 = 2 \cdot 2 \cdot 2 \cdot 2.$$

Proceed as in Example 1:

$$2 . 2 . 2 . 2 . 3 . 5 = 240.$$

∴ the L.C.M. of 8, 12, 15, 16 is 240.

Exercises 2

Find the L.C.M. of the following:

1. 16, 24, 32.   2. 12, 16, 20.
3. 72, 135, 300.   4. 3, 12, 18, 36.
5. 7, 18, 21, 35.   6. 5, 15, 24, 32, 35.

7. Four bells start ringing together at 12 o'clock. They ring at intervals of 2, 6, 8 and 10 sec. How many times will they have rung together by 12.30?

8. Two spur wheels engage with each other. One has 56 teeth and the other 48. After how many revolutions of the smaller wheel will the same two teeth be in contact?

9. The carrier wheel of a bicycle is 16 in. in diameter, and the back wheel is 28 in. The lowest point of both tyres is marked. After how many revolutions will these two marks be on the ground again at the same time?

10. Three lighthouses flash at intervals of 15, 20 and 25 sec. respectively. How many times during an hour are the flashes seen simultaneously?

FRACTIONS

3. If a rod is divided into a number of equal parts, each part is called a **fraction** of the whole. Supposing it is divided into 8 parts, each part is called an eighth and is written $\frac{1}{8}$. Similarly 2 parts is written $\frac{2}{8}$. The figure above the horizontal line is called the **numerator**, and the figure below this line is referred to as the **denominator**. The denominator denotes how many equal parts the unit has been divided into, and the numerator indicates how many of those parts are being considered. The fraction $\frac{5}{16}$ means that some unit has been divided into 16 parts, and that 5 of those parts are being considered.

A **proper fraction** is one where the denominator is greater than the numerator. Thus $\frac{4}{7}$, $\frac{5}{8}$ are proper fractions.

An **improper fraction** is a fraction in which the denominator is less than the numerator, such as $\frac{15}{4}$, $\frac{12}{9}$. The fraction $\frac{15}{4}$ denotes that some unit has been divided into four equal parts, and that a value equivalent to fifteen of these parts is considered.

A **mixed number** consists of a whole number and a fraction; for example $3\frac{3}{5}$, $2\frac{7}{9}$ are mixed numbers.

A unit may be expressed as a fraction, e.g. $\frac{5}{5}$, because this fraction denotes that something is divided into 5 equal parts, and 5 are taken, i.e. the whole is taken.

An improper fraction may be expressed as a mixed number. Thus $\frac{10}{3}=\frac{3}{3}+\frac{3}{3}+\frac{3}{3}+\frac{1}{3}=3\frac{1}{3}$. This result may be obtained by dividing the denominator into the numerator. Conversely, a mixed number may be expressed as an improper fraction by multiplying the whole number by the denominator of the fraction and adding in the numerator. Thus $5\frac{2}{3}=\frac{17}{3}$, $4\frac{3}{8}=\frac{35}{8}$. The reason for this process is easily seen.

Exercises 3

Express the following improper fractions as mixed numbers:

1. $\frac{45}{8}$, $\frac{23}{7}$, $\frac{17}{9}$, $\frac{12}{5}$, $\frac{67}{4}$, $\frac{38}{6}$.

2. $\frac{145}{18}$, $\frac{237}{23}$, $\frac{416}{19}$, $\frac{826}{41}$, $\frac{79}{17}$.

Change the following mixed numbers to improper fractions:

3. $9\frac{3}{8}$, $17\frac{4}{7}$, $6\frac{4}{15}$, $21\frac{7}{12}$, $15\frac{3}{4}$.

4. $10\frac{1}{11}$, $19\frac{4}{13}$, $25\frac{5}{7}$, $14\frac{4}{9}$, $18\frac{7}{10}$.

4. We know that to divide something into two parts and take one of them is the same as dividing it into four parts and taking two of them, or dividing it into eight parts and taking four.

$$\therefore \quad \tfrac{1}{2}=\tfrac{2}{4}=\tfrac{4}{8},$$

viz. $$\frac{1}{2}=\frac{1\times2}{2\times2}=\frac{1\times4}{2\times4},$$

that is, **if we multiply the numerator and denominator of a fraction by the same number, the value of the fraction remains unaltered. The converse is equally true. That is, if the numerator and denominator of a fraction are divided by the same number its value remains unaltered.**

Thus $$\frac{4}{8}=\frac{4\div4}{8\div4}=\frac{1}{2}.$$

When the numerator and denominator are made as small as

possible by dividing both by the same number, the fraction is said to be reduced to its lowest terms.

EXAMPLE. Reduce $\frac{36}{45}$ to its lowest terms.

We see that 36 and 45 are each divisible by 9.

$$\therefore \quad \frac{36}{45} = \frac{4}{5},$$

and, as nothing further will divide into both 4 and 5, $\frac{4}{5}$ is the lowest terms in which to express $\frac{36}{45}$.

Should a common factor not readily be seen, the H.C.F. of numerator and denominator should be found.

Exercises 4

Express the following fractions in their lowest terms, and where necessary as a mixed number:

1. $\frac{15}{63}, \frac{36}{60}, \frac{78}{15}, \frac{42}{189}$.

2. $\frac{144}{132}, \frac{78}{169}, \frac{85}{323}, \frac{253}{92}$.

ADDITION OF FRACTIONS

5. The sum of a number of fractions having the same denominator is found by adding their numerators together, e.g.

$$\frac{2}{9} + \frac{4}{9} + \frac{5}{9} + \frac{8}{9} = \frac{19}{9} = 2\frac{1}{9},$$

just as 2 apples + 4 apples + 5 apples + 8 apples = 19 apples.

When the denominators are not the same, then the fractions cannot thus be added directly.

If the denominators are different they must be made the same before addition can take place.

EXAMPLE 1. $\qquad\qquad \frac{1}{10} + \frac{4}{5}$

cannot be added until both denominators are in the same terms.

Thus $\qquad\qquad\qquad \frac{4}{5} = \frac{8}{10}$.

$$\therefore \quad \frac{1}{10} + \frac{4}{5} = \frac{1}{10} + \frac{8}{10} = \frac{9}{10}.$$

To express a number of fractions having different denominators as having the same denominator, find the L.C.M. of the denominators. The L.C.M. thus found is the number in which all the denominators will be expressed. It is called the **common denominator**.

EXAMPLE 2. Add $\frac{3}{4}, \frac{5}{8}, \frac{7}{12}$.

The L.C.M. of 4, 8, 12 is 24. Therefore all the denominators must be 24.

$$\frac{3}{4} = \frac{18}{24}; \ \frac{5}{8} = \frac{15}{24}; \ \frac{7}{12} = \frac{14}{24}.$$

$$\therefore \quad \frac{3}{4} + \frac{5}{8} + \frac{7}{12} = \frac{18}{24} + \frac{15}{24} + \frac{14}{24} = \frac{47}{24} = 1\frac{23}{24}.$$

EXAMPLE 3. Find the sum of $2\frac{5}{12}, 3\frac{4}{15}, 1\frac{1}{18}$.

The L.C.M. of 12, 15, 18 is 180.

$$\therefore \quad 2\frac{5}{12} + 3\frac{4}{15} + 1\frac{1}{18} = 6 + \frac{75}{180} + \frac{48}{180} + \frac{10}{180} = 6\frac{133}{180}.$$

Exactly the same procedure is followed in subtraction of fractions.

EXAMPLE 4. From $5\frac{5}{12}$ take $2\frac{4}{15}$.

$$5\frac{5}{12} - 2\frac{4}{15} = 3\frac{5}{12} - \frac{4}{15}.$$

The L.C.M. of 12 and 15 is 60.

$$\therefore \quad 3\frac{5}{12} - \frac{4}{15} = 3\frac{25}{60} - \frac{16}{60} = 3\frac{9}{60} = 3\frac{3}{20}.$$

EXAMPLE 5. From $7\frac{5}{18}$ take $3\frac{11}{24}$.

$$7\frac{5}{18} - 3\frac{11}{24} = 4\frac{5}{18} - \frac{11}{24}.$$

The L.C.M. of 18 and 24 is 72.

$$\therefore \quad 4\frac{5}{18} - \frac{11}{24} = 4\frac{20}{72} - \frac{33}{72}.$$

Now we cannot subtract 33 from 20. We therefore have to subtract $\frac{33}{72}$ from one of the units in 4. It may thus be written

$$3 + \frac{72}{72} + \frac{20}{72} - \frac{33}{72} = 3 + \frac{39}{72} + \frac{20}{72} = 3\frac{59}{72}.$$

Exercises 5

1. Arrange in order of magnitude and add the least to the greatest:

$$\frac{1}{9}, \frac{80}{81}, \frac{4}{27}, \frac{11}{12}.$$

2. $\frac{4}{5} + 1\frac{2}{3} + 1\frac{1}{12} - \frac{11}{20}.$

3. Find the overall lengths of the screwed bolts (or spindles) shown in Fig. 1 (a), (b), (c), (d).

4. $(4\frac{3}{8} + 2\frac{1}{5}) - (1\frac{11}{24} + 3\frac{7}{12}).$

5. $1\frac{7}{15} + 1\frac{9}{20} - 2\frac{1}{24}.$

6. $(2\frac{1}{2} + 3\frac{3}{4}) - (4\frac{1}{16} - 3\frac{1}{12}).$

7. $(\frac{5}{14} + 6\frac{1}{2}) - (3\frac{8}{21} + 2\frac{10}{21})$.

8. $5\frac{3}{8} + 7\frac{3}{10} - 1\frac{3}{5}$.

9. $3\frac{8}{9} - (1\frac{3}{14} + \frac{5}{8} + \frac{4}{21})$.

10. By how much does the sum of $7\frac{3}{5}$ and $5\frac{3}{4}$ exceed their difference?

11. $(14\frac{1}{2} + 9\frac{1}{8}) - (25\frac{3}{4} - 16\frac{1}{3})$.

12. $(\frac{3}{10} + \frac{1}{6} + 1\frac{1}{15}) - (\frac{1}{4} + \frac{5}{24} + \frac{1}{6})$.

13. The length of a belt is 40 yd. What length will remain when three pieces $4\frac{7}{8}$ yd., $10\frac{3}{4}$ yd. and $6\frac{5}{8}$ yd. have been cut off?

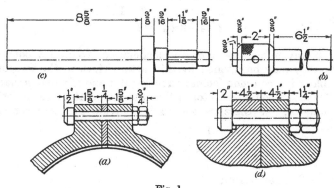

Fig. 1

14. Two towns are $15\frac{1}{4}$ miles apart. Two men start to walk towards each other from these towns. When they have walked $8\frac{3}{8}$ miles and $4\frac{2}{3}$ miles respectively, how far apart are they?

15. In an alloy of zinc, copper and tin three-fourths is copper, $\frac{1}{16}$ is tin and there are 48 lb. of zinc. Find the weight of each metal.

MULTIPLICATION OF FRACTIONS

6. If 3 is multiplied by 4 the result is 12,

written $3 \times 4 = 12$.

If we multiply three-sevenths by 4 we get twelve-sevenths,

written $\frac{3}{7} \times 4 = \frac{12}{7}$.

Therefore, to multiply a fraction by a whole number, we multiply the numerator by that number.

7. To multiply one fraction by another

$$\tfrac{2}{3} \times \tfrac{3}{4}.$$

This means that we require to find the value of $\tfrac{3}{4}$ of $\tfrac{2}{3}$.

Now $\qquad\qquad \tfrac{1}{4}$ of $\tfrac{1}{3} = \tfrac{1}{12}$,

and $\qquad\qquad \tfrac{1}{4}$ of $\tfrac{2}{3} = \tfrac{2}{12}$.

$\qquad\qquad \therefore \quad \tfrac{3}{4}$ of $\tfrac{2}{3} = \tfrac{2}{12} \times 3 = \tfrac{6}{12} = \tfrac{1}{2}$,

that is, the value of $\tfrac{2}{3}$ of $\tfrac{3}{4}$ has been obtained by multiplying the two numerators together for the final numerator, and the two denominators together for the final denominator.

That is, $\qquad\qquad \dfrac{2}{3} \times \dfrac{3}{4} = \dfrac{2 \times 3}{3 \times 4} = \dfrac{6}{12} = \dfrac{1}{2}$.

8.

If any numerator and any denominator contain a common factor they may be divided by this factor. Thus, in the foregoing example, 3 is a common factor of a numerator and of a denominator, and 2 is also a common factor of the other numerator and the other denominator. We therefore successively divide by these common factors:

$$\overset{1}{\underset{1}{\cancel{\tfrac{2}{3}}}} \times \overset{1}{\underset{2}{\cancel{\tfrac{3}{4}}}} = \frac{1 \times 1}{1 \times 2} = \frac{1}{2}.$$

This is called **cancelling**.

The foregoing may also be proved by means of a diagram.

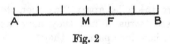

Fig. 2

Draw a line AB 6 units long. Mark off the units.

Now $\qquad\qquad AF = \tfrac{2}{3}$ of AB,

and $\qquad\qquad AM = \tfrac{3}{4}$ of AF,

$\qquad\qquad\qquad = \tfrac{3}{4}$ of $\tfrac{2}{3}$ of AB.

But as shown $\qquad AM = \tfrac{1}{2}$ of AB.

$\qquad\qquad \therefore \quad \tfrac{3}{4}$ of $\tfrac{2}{3} = \tfrac{1}{2}$.

9. To divide one fraction by another fraction.

To divide a fraction by a whole number we simply divide the numerator by that number.

Example: $\frac{10}{4} \div 2 = \frac{5}{4}$; $\frac{15}{3} \div 3 = \frac{5}{3}$.

But exactly the same result will be obtained if we multiply the denominator by the dividing number.

Thus $\qquad \frac{10}{4} \div 2 = \frac{10}{8} = \frac{5}{4}$,

and $\qquad \frac{15}{3} \div 3 = \frac{15}{9} = \frac{5}{3}$.

Now suppose we require to divide $\frac{10}{4}$ by $\frac{5}{2}$. If we divide by 5, by multiplying the denominator 4 by 5, we get a result only half what it should be,

viz. $\qquad \frac{10}{4} \div 5 = \frac{10}{20}$.

In order to get the correct result we shall have to multiply $\frac{10}{20}$ by 2. So now what really has happened is that the original fraction which was to be divided by $\frac{5}{2}$ has been multiplied by $\frac{2}{5}$. From this we get the general rule for dividing a fraction by a fraction. **To divide a fraction by a fraction invert the divisor and multiply by the fraction thus obtained.**

EXAMPLE 1. $\qquad \dfrac{5}{8} \div \dfrac{7}{16} = \dfrac{5}{\underset{1}{8}} \times \dfrac{\overset{2}{16}}{7} = \dfrac{10}{7} = 1\dfrac{3}{7}$.

In all questions of division and multiplication of fractions mixed numbers must be expressed as improper fractions.

EXAMPLE 2.
$$2\frac{1}{7} \div 2\frac{13}{21}$$
$$= \frac{15}{7} \div \frac{55}{21}$$
$$= \frac{\overset{3}{15}}{\underset{1}{7}} \times \frac{\overset{3}{21}}{\underset{11}{55}}$$
$$= \frac{9}{11}.$$

Exercises 6

1. $\frac{7}{16} \times \frac{32}{35} \times \frac{5}{8}$.

2. $3\frac{1}{2} \times \frac{4}{7} \times 3\frac{1}{16}$.

3. $2\frac{1}{4} \times 18\frac{2}{3} \times 2\frac{5}{8} \times \frac{1}{18}$.

4. $(1\frac{2}{3} + \frac{4}{5} - 1\frac{8}{9}) \times (\frac{1}{4} - \frac{1}{13})$.

5. $(\frac{2}{3} + \frac{3}{5}) \times (1\frac{2}{3} - \frac{5}{12})$.

6. $2\frac{2}{3} \div 24\frac{8}{9}$.

7. $\frac{11}{20} \div \frac{11}{18}$.

8. $\frac{9}{64} \div (\frac{5}{8} \times \frac{3}{5})$.

9. $19\frac{3}{5} \times 4\frac{1}{6} \div 4\frac{1}{12}$.

10. $4\frac{3}{10} \div (7\frac{1}{6} \times \frac{4}{5})$.

10. When a number of fractions are joined together by different signs, those connected by the signs of multiplication or division are evaluated before those connected by the signs of addition or subtraction, **that is, the signs of multiplication and division bind more closely than the signs of addition and subtraction.**

However, should any of the fractions be enclosed in brackets, then these fractions inside the brackets must first be evaluated no matter what sign connects the brackets to another fraction.

EXAMPLE 1.
$$\frac{3}{4} \times \frac{16}{21} - \frac{3}{14}.$$

$$\frac{3}{4} \times \frac{16}{21} = \frac{4}{7}.$$

$$\therefore \quad \frac{4}{7} - \frac{3}{14} = \frac{5}{4} = 1\frac{1}{4}.$$

EXAMPLE 2.
$$2\frac{1}{4} + 3\frac{1}{5} \div \frac{12}{25}.$$

$$3\frac{1}{5} \div \frac{12}{25} = \frac{16}{5} \times \frac{25}{12} = \frac{20}{3} = 6\frac{2}{3}.$$

$$\therefore \quad 2\frac{1}{4} + 6\frac{2}{3} = 8\frac{11}{12}.$$

EXAMPLE 3.
$$\left(\frac{2}{3} + 4\frac{7}{12}\right) \times \frac{12}{35} - \frac{7}{10}.$$

$$\frac{2}{3} + 4\frac{7}{12} = 5\frac{1}{4}.$$

$$5\frac{1}{4} \times \frac{12}{35} = \frac{21}{4} \times \frac{12}{35} = \frac{9}{5} = 1\frac{4}{5}.$$

$$\therefore \quad 1\frac{4}{5} - \frac{7}{10} = 1\frac{1}{10}.$$

Exercises 7

1. $\frac{2}{3} + \frac{3}{5}$ of $1\frac{2}{3} - \frac{5}{12}.$

2. $7\frac{1}{2} - 3\frac{1}{3} \div \frac{8}{15}.$

3. $\frac{5}{8}$ of $\frac{14}{15} - \frac{3}{4}$ of $\frac{5}{9}.$

4. $\frac{3}{7} \div 1\frac{13}{14} + 3\frac{1}{8} \times \frac{24}{35}.$

5. $2\frac{1}{4} \times (3\frac{1}{2} - \frac{7}{8}) \div 1\frac{3}{4}$.

6. $(\frac{7}{8} - \frac{1}{3}$ of $\frac{9}{30}) \times (4\frac{1}{17} - 3\frac{14}{51})$.

7. $7\frac{1}{8} + 3\frac{1}{8} \times \frac{4}{25} - 4\frac{1}{5} \div 1\frac{3}{25}$.

8. $\frac{15}{16} \times 3\frac{5}{9} - 1\frac{4}{5} \div 1\frac{3}{5}$.

9. $\frac{5}{16} \div 1\frac{7}{8} + 1\frac{1}{3}$ of $1\frac{4}{5}$.

10. $(2\frac{1}{8} \times 2\frac{4}{19} - 1\frac{7}{8} \times 2\frac{2}{5}) \div 4\frac{1}{19}$.

11. When a series of fractions are separated from another series by a horizontal line, the value of the expression above the line is divided by the value below the line.

EXAMPLE.
$$\frac{3\frac{1}{8} \text{ of } \frac{4}{25}}{2\frac{1}{3} + \frac{1}{2}}.$$

$$3\frac{1}{8} \text{ of } \frac{4}{25} \qquad\qquad 2\frac{1}{3} + \frac{1}{2}$$
$$= \frac{25}{8} \times \frac{4}{25} = \frac{1}{2}. \qquad = 2\frac{5}{6}.$$

$$\therefore \quad \frac{1}{2} \div 2\frac{5}{6} = \frac{1}{2} \div \frac{17}{6} = \frac{1}{2} \times \frac{\overset{3}{\cancel{6}}}{17} = \frac{3}{17}.$$

Exercises 8

1. $\dfrac{3\frac{1}{17} + 17\frac{1}{3}}{\frac{1}{17} \text{ of } 2\frac{1}{6}}$.

2. $\dfrac{\frac{1}{2} + \frac{3}{5} + \frac{5}{7}}{2\frac{1}{2} + \frac{4}{5} + \frac{1}{7}}$.

3. $\dfrac{\frac{1}{2} + \frac{1}{3} - \frac{1}{6}}{\frac{1}{2} \text{ of } \frac{1}{3} \text{ of } \frac{1}{6}}$.

4. $\dfrac{3\frac{1}{2} - 2\frac{1}{4} + 1}{3\frac{1}{2} + 2\frac{1}{4} - 1}$.

5. $\dfrac{2\frac{1}{4} \text{ of } \frac{8}{15} - \frac{1}{2}}{\frac{1}{2} + \frac{1}{3} \times \frac{1}{15}}$.

6. $\dfrac{\frac{5}{6} + \frac{4}{9} \text{ of } 1\frac{1}{8} - \frac{3}{4}}{\frac{3}{8} \text{ of } 1\frac{1}{3} + 2\frac{1}{2} \div 2\frac{1}{12}}$.

7. $2 - \dfrac{\frac{3}{5}}{2\frac{1}{2} \times 1\frac{1}{5}}$.

8. $\dfrac{1 + \frac{1}{3} - \frac{1}{2} - \frac{1}{8}}{1 - \frac{1}{3} + \frac{1}{2} + \frac{1}{8}}$.

9. $\dfrac{\frac{4}{5} + \frac{2}{3}}{\frac{4}{5} - \frac{2}{3}} \times \dfrac{12}{3\frac{1}{7}}$.

10. $\frac{1}{13}$ of $1\frac{5}{8}$ of $\dfrac{\frac{1}{3} + \frac{1}{4} + \frac{1}{12}}{\frac{1}{6} + \frac{1}{8} + \frac{3}{4}}$.

12. To express one quantity as the fraction of another quantity.

If necessary, first express both quantities in the same denomination, then make the quantity which has to be expressed as the fraction of the other the numerator, and the other quantity the denominator.

EXAMPLE 1. Express 2s. 6d. as a fraction of 10s.

Thus   2s. 6d. = 5 sixpences,

and   10s. = 20   ,,

$\therefore$ the fraction is $\frac{5}{20} = \frac{1}{4}$,

that is, 2s. 6d. is $\frac{1}{4}$ of 10s.

EXAMPLE 2. Express 64 yards as a fraction of 1 mile.

1 mile = 1760 yd.

$\therefore$ fraction is $\frac{64}{1760} = \frac{2}{55}$,

that is, 64 yd. are $\frac{2}{55}$ths of a mile.

Note. *An easy working rule is to make denominator the quantity preceded by "of".*

EXAMPLE 3. What fraction of £1. 10s. 11d. is 13s. 3d.?

13s. 3d. = 159d.

£1. 10s. 11d. = 371d.

$\therefore$ fraction is $\frac{159}{371}$.

The H.C.F. of numerator and denominator is 53.

$\therefore \frac{159}{371} = \frac{3}{7}$.

Exercises 9

1. Express 3s. 8d. as a fraction of 11s.

2. What fraction of 17 yd. 1 ft. is 6 ft. 6 in.?

3. Express one-half of £2. 13s. 4d. as the fraction of $\frac{1}{6}$ of £11.

4. Five boys earn 3s. 6d. each a day, and four girls earn 2s. 6d. each. Express the girls' wages as a fraction of the boys'.

5. Reduce 5 cwt. 2 qr. 14 lb. to a fraction of 14 cwt. 0 qr. 7 lb.

6. Express 3 gall. 1 qt. 1 pt. as a fraction of 14 gall. 2 qt. 1 pt.

7. What fraction of £6. 19s. 5d. is $\frac{2}{3}$ of $\frac{9}{10}$ of £4. 2s. 11d.?

8. What fraction of 2 tons 11 cwt. 1 qr. 2 lb. is $\frac{5}{28}$ of a ton $+ \frac{5}{16}$ of a cwt.?

9. What sum of money is the same fraction of £3. 2s. 6d. as £1. 3s. 3d. is of £9. 13s. 9d.?

10. What length is the same fraction of $\frac{1}{2}$ mile as £3. 3s. 3d. is of £21. 1s. 8d.?

13. Examination of Regular Solids.

1. Fig. 3 shows certain well-known geometrical solids—models of which should be accessible to the student.

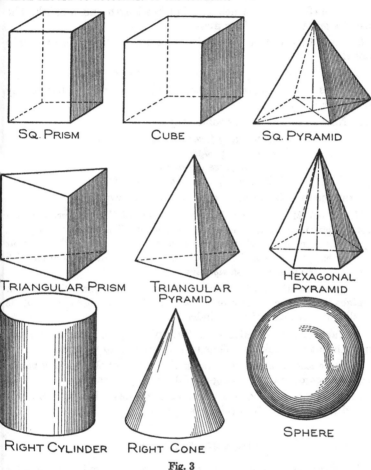

SQ. PRISM   CUBE   SQ. PYRAMID

TRIANGULAR PRISM   TRIANGULAR PYRAMID   HEXAGONAL PYRAMID

RIGHT CYLINDER   RIGHT CONE   SPHERE

Fig. 3

Firstly observe that all "*occupy space*" or "*take up room*"; they are 3-dimensional, i.e. they extend in three directions—having length, breadth, and height. The size of a solid is called its **volume**.

2. Take a typical solid such as the square prism. Notice the following:

(a) The faces, or **surfaces**, are **flat**.

(b) A surface has no thickness, but has length and breadth. The size of a surface is called its **area**.

(c) Any two surfaces meet in an **edge** or **line**. **Lines have length**.

(d) Any two or more edges meet in a corner or **point**.

(e) **Points** have neither length, breadth, nor height, but simply **mark position**.

3. Now consider another solid—say a cube.

> *How many faces has it?*
> *How many edges?*
> *Are the faces flat or curved?*
> *Are the edges straight or curved?*

Examine models of other solids in the same way.

Hints. (a) Mark two small crosses on one of the flat surfaces of a rectangular prism. The intersection of two fine lines indicates a point. Apply the edge of a steel rule so as to *join the two points*. Observe that no light is seen beneath the edge of the rule, i.e. it makes contact with the face of the prism in *a straight line*.

(b) Repeat this on the curved surface of a cylinder. Observe that straight line contact is now only possible when the edge of the rule is parallel with the axis of the cylinder.

4. *Summarising.* A **plane** is a **flat surface**, i.e. it has length and breadth but no thickness. Thus it has area but no volume.

For practical purposes we consider the surface of the drawing paper, the blackboard, or the marking-off slab as plane surfaces.

Lines, in geometry, are considered to have **length but not breadth**. They may be straight or curved.

PLANE FIGURES

14. Examine the bounding plane surfaces of the cube, rectangular prism, pyramid, cylinder, cone, etc. shown in Fig. 3. We name each of these surfaces according to (1) the number of lines bounding it, (2) its shape. Let us enumerate a few plane figures (see Fig. 4).

Bounded by one line :   *circle, ellipse, oval.*[1]
  ,,   ,, *two lines:*   *segments of circles.*
  ,,   ,, *three ,, :*   *sectors of circles, triangles.*
  ,,   ,, *four ,, :*   *quadrilaterals.*
  ,,   ,, *five ,, :*   *pentagons.*
  ,,   ,, *six ,, :*   *hexagons.*
  ,,   ,, *seven ,, :*   *heptagons.*
  ,,   ,, *eight ,, :*   *octagons.*
  ,,   ,, *nine ,, :*   *nonagons.*
  ,,   ,, *ten ,, :*   *decagons.*

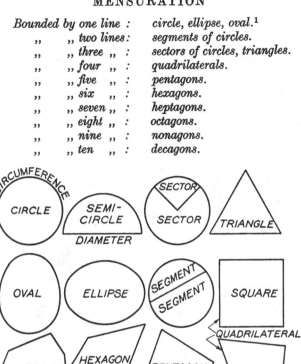

Fig. 4

Learn these definitions:

1. **Plane figures** are portions of plane or flat surfaces bounded by one or more lines.

2. **Rectilinear figures** are bounded by straight lines.

They are:

Equilateral when all sides are equal;

Equi-angular when all angles are equal;

Regular when both equilateral and equi-angular.

---

[1] Latin, *ovum*: an egg.

Exercises 10

Oral and Class-work (Fig. 5)

1. What comprises the shortest distance between two points?

2. Endeavour to draw a straight line on the surface of a cube and on a cone and cylinder.

3. Mark any plane surfaces you can identify on a cone, cylinder, cube, hexagonal pyramid, and sphere.

SOME INTERESTING SECTIONS

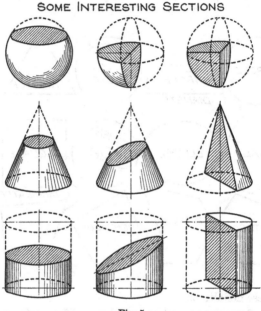

Fig. 5

4. A sphere is imagined as cut by a plane. What is the true shape of the resulting section?

5. A sphere is cut into four equal parts by two straight cuts. How many surfaces has each resulting part? How many of these surfaces are plane? How many edges are straight?

6. A cutting plane is parallel to the base of a right cone. What is the true shape of the section?

7. Suppose the cutting plane in Question 6 passed through the axis of the right cone at right angles to its base. What would then be the true shape of the section?

8. What surfaces of a right cylinder lie in parallel planes?

9. How would you cut a cylinder so as to produce a plane section bounded by (a) one line, (b) four straight lines?

10. A stiff piece of paper or cardboard is folded. Is the "crease" straight or curved? Why?

PRELIMINARY MENSURATION

15. The **area** of a plane figure is the amount of space contained within its bounding lines.

The **perimeter** of a plane figure is the total length of its bounding lines.

Thus the perimeter of the rectangle $ABCD$ (Fig. 6)

$$= AB + BC + CD + DA$$
$$= 3 + 4 + 3 + 4 \text{ units}$$
$$= 14 \text{ units}.$$

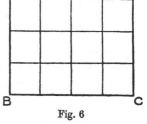

Fig. 6

Lengths of lines or edges are expressed in *linear units*.

Areas of surfaces are expressed in *square units*.

We can readily apply our work in multiplication, fractions, etc. to the calculation of simple areas and volumes.

Thus consider the rectangle $ABCD$ in Fig. 6. Its linear dimensions are evidently 4 units by 3 units. Let us suppose $AB = 3$ in., $BC = 4$ in. Divide AB into 3 equal parts and BC into 4 equal parts. Through the points obtained draw the light parallel lines, thus dividing the figure into 3 strips each containing 4 sq. units, i.e. in this case, 4 sq. in. Then area of $ABCD = 4 \times 3 = 12$ sq. in.

Notice that this result is obtained by multiplying the number of units in AB by the number of units in BC. It can readily be shown that this result also applies where fractions of units occur in length and breadth. Thus for any rectangle,

Area = length × breadth.

Note the length and breadth must be expressed in terms of the same linear unit. The area will then be expressed in corresponding square units.

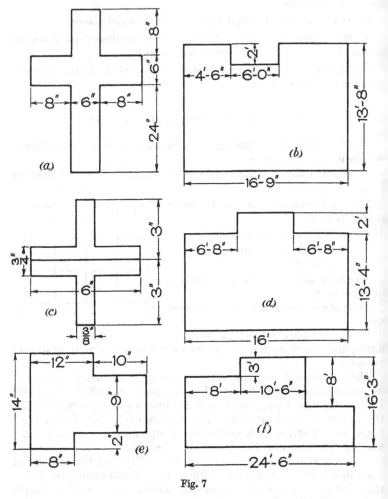

Fig. 7

EXAMPLE. The dimensions of a rectangle are 4 ft. 6 in. by 2 yd. 3 in. Calculate (a) its area, (b) its perimeter.

(a) $A = l \times b$

$\qquad = 4\frac{1}{2} \times 6\frac{1}{4}$ (expressing both length and breadth in ft.)

$\qquad = \frac{9}{2} \times \frac{25}{4}$

$\qquad = \frac{225}{8} = 28\frac{1}{8}$ sq. ft.

(b) $P = 2\,(l+b)$

$\qquad = 2\,(4\frac{1}{2} + 6\frac{1}{4})$

$\qquad = 2\,(10\frac{3}{4}) = 21\frac{1}{2}$ ft.

Exercises 11

1. Find the area of a floor 8 ft. 6 in. by 6 ft. 6 in.

2. The sizes of a room are: length 12 ft., breadth 10 ft., height 8 ft.

$\quad$ (a) What length of picture rail is required to go all round the walls?

$\quad$ (b) What is the area of the ceiling?

$\quad$ (c) What is the area of the four walls (neglecting doors and windows)?

3. The cross-sectional dimensions of an underground railway are 8 ft. 6 in. by 5 ft. 9 in. Find the area of the cross-section.

4. Find the area of the cross shown in Fig. 7 (a).

5. Fig. 7 (b) shows the plan of a shop. Find its area in square feet. It is proposed to nail a frieze all round the walls. How long will it be?

6. Fig. 7 (c) shows the section of a column used to support a beam. Find its area in square inches.

7. Fig. 7 (d) shows the plan of a room. Find (a) its area in square feet, (b) the length of a frieze to go right round its walls.

8. Find the area and perimeter of the plane figure shown in Fig. 7 (e).

9. Find the area and perimeter of the octagonal area shown in Fig. 7 (f).

THE AREA OF A TRIANGLE

16. Take a rectangular sheet of paper similar to Fig. 8 ($ABCD$). Draw a diagonal BD. Cut along BD and so obtain two triangles ABD and BDC. Compare their areas. They are equal.

$\qquad$ Area of rect. $ABCD = $ base $\times$ height.

$\therefore \qquad ,, \qquad \triangle \qquad BDC = \frac{1}{2}$ (**base** $\times$ **height**).

Now consider the $\triangle EFD$. Later we shall prove that:

$$\text{Area of } \triangle EFD = \tfrac{1}{2} \text{ (base} \times \text{height)}$$
$$= \tfrac{1}{2} (EF \times GD).$$

Note that by "height" we mean *perpendicular height*, i.e. the length of the perpendicular from D to the base or the base produced.

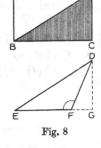

Fig. 8

EXAMPLE 1. Find the area of the triangle shown in Fig. 9 (a).

$$\text{Area} = \tfrac{1}{2} \text{ (base} \times \text{height)}$$
$$= \tfrac{1}{2} (20 \times 12)$$
$$= 120 \text{ sq. ft.}$$

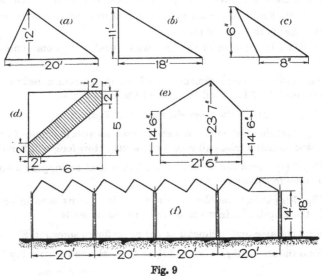

Fig. 9

EXAMPLE 2. Find the area of the shaded part of the rectangle shown in Fig. 9 (d).

The area can be found by subtracting the areas of the two unshaded triangles from the area of the rectangle. Suppose dimensions are in cm.

$$\text{Area of rectangle} \qquad\quad = 6 \times 5 \qquad = 30 \text{ sq. cm.}$$

$$\text{Area of unshaded triangles} = 2 \left(\frac{4 \times 3}{2} \right) = 12 \qquad \text{,,}$$

$$\text{Area of shaded figure} \qquad = 30 - 12 \quad = 18 \qquad \text{,,}$$

1. Find the areas of the triangles shown in Fig. 9 (*b*), (*c*).

2. Suppose Fig. 9 (*a*) is the plan of a triangular yard. Find the cost of tar paving it at 5*s*. 6*d*. per sq. yd.

3. The cross-section of a workshop is shown in Fig. 9 (*e*). Find the area of this cross-section in sq. ft.

4. In Fig. 9 (*f*) we show the section of a saw-tooth roofed workshop. Find its cross-sectional area.

DECIMALS

17. A figure in any number depends for its value on the position it holds in the number. This is called **place value**. The number 111 reads one hundred and eleven, that is 100 units + 10 units + 1 unit, each figure being ten times the value of the figure on its right. Now if we place another figure 1 to the right of the number 111 it will represent only the tenth part of a unit. To mark the boundary between figures representing *whole units* and figures representing *fractions of a unit* a mark is placed immediately after the last figure representing whole units. This mark is called the **decimal point** (Lat. *decem*, ten). As each figure to the right occupies a place one tenth the value of the place preceding it, the first figure after the decimal point will represent tenths, the next figure the tenth of a tenth, namely one hundredth, and so on.

EXAMPLE 1. The number 3·64 represents 3 whole units, 6 tenths of a unit and 4 hundredths of a unit.

EXAMPLE 2. If the diameter of a screw gauge as measured by a micrometer is 2·6475 in., give each figure its place value.

$$2\cdot0000 = 2 \text{ in.}$$
$$\cdot6000 = 6 \text{ tenths of an in.}$$
$$\cdot0400 = 4 \text{ hundredths of an in.}$$
$$\cdot0070 = 7 \text{ thousandths of an in.}$$
$$\cdot0005 = 5 \text{ ten thousandths of an in.}$$
$$\overline{2\cdot6475}$$

18. As the figures to the right of the decimal point represent fractions, ·6475 is called a decimal fraction.

19. Addition and subtraction of decimals.

This is performed in the same way as in the case of whole numbers, care being taken that figures of the same place value are added or subtracted.

20. Powers of 10.

When 10 is multiplied by itself any number of times, we obtain a result called a *power of* 10. Thus $10 \times 10 = 100$. The result 100 is called 10 to the second power and may be written 10^2. Similarly $10 \times 10 \times 10 = 1000$, and the result is called 10 to the power of 3, and may be written 10^3.

21. To multiply or divide by powers of 10.

EXAMPLE 1.
$$2 \cdot 394 \times 10 \quad = 23 \cdot 94,$$
$$2 \cdot 394 \times 100 \quad = 239 \cdot 4,$$
$$2 \cdot 394 \times 1000 = 2394 \cdot.$$

It will be noticed that the decimal point moves as many places **to the right** as there are ciphers in the multiplier. The converse is true for division, the decimal point moving as many places **to the left** as there are ciphers in the divisor.

EXAMPLE 2.
$$567 \cdot 3 \div 100 = 5 \cdot 673,$$
$$5 \cdot 673 \div 100 = \cdot 05673.$$

In the latter division we have to move the decimal point 2 places to the left. That will bring the decimal point to the left of the 5 and a place further. This further place has to be filled by a cipher as shown.

22. Multiplication by a whole number.

EXAMPLE.
$$4 \cdot 26 \times 7 = 29 \cdot 82.$$

6 hundredths $\times 7 = 42$ hundredths $= 4$ tenths $+ 2$ hundredths. The 2 hundredths is put down in the answer, then the 4 tenths is added on to the product of 2 tenths and 7, and so on until the multiplication is complete.

23. Multiplication by a decimal.

EXAMPLE 1. Suppose instead of multiplying $4 \cdot 26$ by 7 we require to multiply it by $\cdot 7$. As $\cdot 7$ is one tenth of 7, the result of multiplying by $\cdot 7$ will be one tenth the result of multiplying by 7.

$$\therefore \quad 4 \cdot 26 \times \cdot 7 = \tfrac{1}{10} \text{ of } 4 \cdot 26 \times 7 = 2 \cdot 982.$$

Another way of looking at this is as follows:

EXAMPLE 2. $4 \cdot 26 \times \cdot 7$.

We require to find the value of 7 tenths of 4·26.

$$1 \text{ tenth of } 4 \cdot 26 = \cdot 426.$$
$$\therefore \quad 7 \text{ tenths} = \cdot 426 \times 7 = 2 \cdot 982.$$

EXAMPLE 3. $2 \cdot 41 \times \cdot 26$.

Here we require to find the value of 26 hundredths of 2·41.

$$1 \text{ hundredth of } 2 \cdot 41 = \cdot 0241.$$
$$\therefore \quad 26 \text{ hundredths} = \cdot 0241 \times 26$$
$$= \cdot 0241 \times 20 + \cdot 0241 \times 6$$
$$= \cdot 482 \qquad + \cdot 1446$$
$$= \cdot 6266.$$

It will be noticed in the three examples given that the number of decimal places in the result is the same as the number of decimal places in the multiplier and multiplicand added together. From this we may state a working rule for multiplication of decimals.

Consider the numbers to be multiplied together as whole numbers. When the product is obtained, place the decimal point as many places from the right as there are decimals to the right in the multiplier and multiplicand added together.

EXAMPLE 4.

$$\begin{array}{r} 5 \cdot 6\,2 \\ 4 \cdot 2\,3 \\ \hline 1\,6\,8\,6 \\ 1\,1\,2\,4 \\ 2\,2\,4\,8 \\ \hline 2\,3 \cdot 7\,7\,2\,6 \end{array}$$

Applying the rule, as there are 4 decimal places in the multiplier and multiplicand together, the decimal point will be placed between the 3 and 7, thus giving 4 decimal places.

Exercises 13

1. $57 \cdot 31 \times 5 \cdot 23$.
2. $12 \cdot 34 \times \cdot 032$.
3. $227 \cdot 8 \times \cdot 3175$.
4. $\cdot 0765 \times 25 \cdot 6$.
5. $\cdot 536 \times \cdot 53$.
6. $4 \cdot 376 \times \cdot 0023$.
7. $\cdot 00715 \times 243 \cdot 5$.
8. $35 \cdot 73 \times \cdot 523$.
9. $\cdot 048 \times 7 \cdot 34$.
10. $43 \cdot 2 \times \cdot 056$.
11. $538 \cdot 5 \times \cdot 016$.
12. $23 \cdot 72 \times 51 \cdot 2$.
13. $35 \cdot 4 \times \cdot 056$.
14. $234 \cdot 5 \times \cdot 521$.

24. Approximations and significant figures.

It may be required to find the product of two or more numbers, correct to (a) the nearest unit, (b) nearest tenth, (c) nearest hundredth, and so on.

When such an approximate result is required, notice must always be taken of the figure immediately to the right of the approximate value required. If this figure is 5 *or over* then the figure to its left is increased by 1. Thus in *Example* 4 (para. 23) the result *to the nearest unit* is 24, as 23·7 is nearer 24 than it is 23. The result *to the nearest tenth* is 23·8, because 23·77 is nearer 23·8 than it is 23·7.

If the figure immediately to the right of the approximate figure is *less than* 5, no alteration is made in the figure on its immediate left. Thus the approximate value of 5·263 *to the nearest hundredth* is 5·26.

In the number ·0034 we have a number of two *significant figures*— the ciphers only fix the position of these figures.

EXAMPLE. Find the product of ·0037 and ·042 to two significant figures.

$$
\begin{array}{r}
·0\ 0\ 3\ 7 \\
·0\ 4\ 2 \\
\hline
7\ 4 \\
1\ 4\ 8 \\
\hline
·0\ 0\ 0\ 1\ 5\ 5\ 4 \\
\hline
\end{array}
$$

The answer correct to two significant figures is ·00016.

Exercises 13 (*contd.*)

15. ·3672 × ·124 to 3 significant figures.
16. ·537 × ·065 to 3 decimal places.
17. 3·872 × ·014 to 2 significant figures.
18. ·025 × ·025 to 2 significant figures.
19. 1·724 × ·015 × ·2 to 3 decimal places.
20. ·04 × ·04 × ·04 to 5 decimal places.

DIVISION OF DECIMALS

25. If the divisor is a whole number the division is done as in ordinary long division, care being taken to place the decimal point in the quotient immediately after the first decimal figure is brought down.

EXAMPLE 1. $77 \cdot 76 \div 24.$

```
          3·24
  24 ) 77·76
        72
        ──
        57   (Approx. result = 78/24 = 3 + .)
        48
        ──
        96
        96    Ans. 3·24.
        ──
```

26. If the divisor contains a decimal, the decimal point is moved as many places to the right as will make the divisor a whole number. At the same time the decimal point in the dividend must be moved an equal number of places. That is, the divisor and the dividend are *both* multiplied by the *same* number, which of course does not affect the value of the result.

EXAMPLE 2. $9 \cdot 81404 \div 2 \cdot 14$
$= 981 \cdot 404 \div 214.$

```
             4·586
  214 ) 981·404
         856
         ───
         1254   (Approx. result = 981/214 = 4 + .)
         1070
         ────
         1840
         1712
         ────
         1284
         1284    Ans. 4·586.
         ────
```

27. The same procedure in approximate results and significant figures applies in division as in multiplication.

Exercises 14

1. $64 \cdot 8 \div 18.$
2. $19 \cdot 95 \div 1 \cdot 05.$
3. $1 \cdot 44312 \div \cdot 017$ correct to 2 decimal places.
4. $363 \div 2 \cdot 65$ correct to the nearest unit.
5. $22 \cdot 97 \div 3 \cdot 16$ correct to 1 decimal place.

6. $165\cdot44 \div 7\cdot282$ correct to 2 decimal places.   **7.** $6\cdot678 \div \cdot252$.

8. $22\cdot876 \div 7\cdot6$.   **9.** $\cdot00384 \div \cdot00032$.

10. $22\cdot876 \div \cdot00076$.   **11.** $\cdot11438 \div 38$.

12. $3\cdot0975 \div 3\cdot75$.   **13.** $\cdot00405 \div 81$.

14. $2\cdot258 \div 1\cdot255$ correct to the nearest tenth.

Evaluate the following:

15. $\dfrac{\cdot003 \times \cdot004}{\cdot006}$.   **16.** $\dfrac{41\cdot82 \times 166\cdot8}{6\cdot273}$.

17. $\dfrac{\frac{4}{5} \text{ of } 5\cdot625}{\frac{2}{3} \text{ of } 17\cdot106}$ to the nearest unit.   **18.** $\dfrac{\cdot026 \times 1\cdot87 \times \cdot0042}{\cdot7 \times 6\cdot63 \times 5\cdot5}$.

19. $\dfrac{\cdot008}{\cdot05} \div \dfrac{62}{\cdot0031}$.   **20.** $\dfrac{\cdot34 \div \cdot034}{\cdot0034 \div 340}$.

28. To express one quantity as the decimal of another quantity. This may be done in two ways.

EXAMPLE 1. Express $2s.$ $3d.$ as the decimal of £1.

$$3d. \div 12 = \cdot25 \text{ shilling.}$$
$$2\cdot25s. \div 20 = \text{£}\cdot1125.$$

The second method is to express one quantity as a fraction of the other quantity and then express this fraction as a decimal fraction. This is done by dividing the denominator into the numerator.

EXAMPLE 2. Express $2s.$ $10d.$ as a decimal of £1. $5s.$ $0d.$

$$2s. \ 10d. = \ 34d.$$
$$\text{£}1. \ 5s. \ 0d. = 300d.$$
$$\therefore \quad \text{fraction is } \frac{34}{300} = \frac{\cdot34}{3} = \cdot133.$$

EXAMPLE 3. What decimal of 3 cwt. 1 qr. 6 lb. is 3 qr. 21 lb.?

$$3 \text{ qr. } 21 \text{ lb.} = 105 \text{ lb.}$$
$$3 \text{ cwt. } 1 \text{ qr. } 6 \text{ lb.} = 370 \text{ lb.}$$
$$\therefore \quad \text{fraction is } \tfrac{105}{370} = \tfrac{7}{25}.$$

$$
\begin{array}{r}
\cdot28 \\
\hline
25 \overline{)\ 70} \\
50 \\
\hline
200 \\
200 \\
\hline
\end{array}
\qquad Ans. \ \cdot28.
$$

29. To change a decimal quantity expressed in one denomination to a lower denomination.

EXAMPLE 1. Express £3·0625 in £ s. d.

$$£·0625$$
$$\underline{20}$$
$$1·2500 \text{ shillings}$$
$$\underline{12}$$
$$3·0000 \text{ pence.} \qquad Ans. \text{ £3. 1s. 3d.}$$

EXAMPLE 2. Express ·4635 mile in yards, feet and inches.

$$·4635 \text{ mile}$$
$$\underline{1760}$$
$$278100$$
$$32445$$
$$4635$$
$$\overline{815·7600} \text{ yd.}$$
$$\underline{3}$$
$$2·28 \text{ ft.}$$
$$\underline{12}$$
$$3·36 \text{ in.}$$

Ans. 815 yd. 2 ft. 3·36 in.

Exercises 15

Reduce the following to the decimal of £1 correct to five places of decimals:

1. 10s. 8d.   **2.** 15s. 6¼d.   **3.** 13s. 11d.   **4.** 6s. 5½d.   **5.** 11s. 4½d.
6. 2s. 11¾d.   **7.** 7s. 9d.   **8.** 6¾d.   **9.** 3s. 10½d.   **10.** 18s. 6¾d.

Find the value of the following:

11. £0·42857 to the nearest farthing.
12. £0·09853 to the nearest farthing.
13. £3·29764 to the nearest farthing.
14. 3·86274 tons to the nearest lb.
15. 5·63827 cwt. to the nearest lb.
16. 0·79326 yd. to the nearest in.
17. 0·38764 mile to the nearest yd.
18. 0·6735 mile to the nearest in.
19. Express 82 ft. 5 in. as a decimal of 1 mile.
20. Express 363 sq. yd. as a decimal of 1 acre.

Revision Exercises

1. If the distance between two consecutive threads of a screw is ·03125 in., how many threads are there in a screw 1·5 in. long?

2. A bicycle travels 2·443 yd. for each turn of the wheels. How many complete turns will the wheels make in travelling a mile?

3. One half of a pole is painted red and $\frac{1}{4}$ of the remainder blue, what fraction is left to paint?

4. If, in the last question, the unpainted part is 1 ft. 6 in. long, find the total length of the pole.

5. Add £$8\frac{1}{12}$ to the sum of £$3\frac{2}{15}$ and £$6\frac{5}{24}$. Give your answer in £ s. d.

6. How often are $2\frac{3}{20}$ ft. contained in $2\frac{3}{20}$ yd.?

7. One person owns $\frac{1}{8}$ of a property, and another person $\frac{1}{3}$. The difference in the values of their shares is £27. Find the value of the whole property.

8. What are the prime factors of 157,542?

9. Find the H.C.F. of 182, 70, 42.

10. Two bars of steel weighing respectively 1 ton 13 cwt. and 2 tons 2 cwt. are made into two girders of equal weight. What is the weight of the largest possible girder? Give your result in lb.

11. Three cisterns contain respectively $73\frac{1}{8}$ gall., $40\frac{5}{8}$ gall. and $65\frac{5}{8}$ gall. Find the capacity of the largest vessel by which the cisterns may be exactly measured.

12. What is the smallest number which, when divided by either 15, 21 or 27, will always have a remainder of 11?

13. Telegraph poles are placed 66 yd. apart. A man walking along a road comes to a pole and a milestone together. How far will he have to walk before he comes to another pole and milestone together?

14. Find the value of $\frac{7}{15}$ of £2. 15s. $7\frac{1}{2}d.$

15. Find the value of $\dfrac{3\frac{1}{4} \times 4\frac{1}{3}}{3\frac{1}{4} - 2\frac{1}{6}}$ of £5.

16. $5\frac{1}{4} - 2\frac{1}{3} + \frac{1}{3}$ of $5\frac{1}{2}$.

17. A quantity of sticks may be made into an exact number of bundles of 24 in a bundle. If they are made into bundles of 11 or of 13 there is 1 over in each case. Find the least number of sticks.

18. What sum of money is the same fraction of £5. 7s. 7½d. as 15 lb. is of $\frac{3}{5}$ of 1 cwt. 3 qr. 14 lb.?

19. How many pieces of wire each $3\frac{3}{8}$ in. long may be cut from a length of 20 yd., and what length will remain over?

20. Three circles have circumferences of 66, 110 and 165 yd. respectively. Three men facing due East start walking round the circles. How many rounds will each have made before they again face due East simultaneously?

21. Find the length in yards of the shortest piece of wire which may be cut into exact lengths of either 12, 18, 27 or 45 in.

22. A barrel is two-thirds full; after $17\frac{1}{2}$ gall. had been withdrawn it was found to be half full. How much does it hold when full?

23. Evaluate $21\cdot7 \times \cdot025 \div \cdot031$.

24. What is the greatest number which will exactly divide into 102, 153, 187?

25. Simplify $\dfrac{\frac{3}{4}+\frac{2}{5}}{\frac{3}{4}-\frac{2}{5}} \times \dfrac{2\frac{2}{3}}{\frac{2}{3}+\frac{3}{7}}$.

26. Add together $\frac{3}{8}$ of 5s. 6d., $\frac{5}{14}$ of $4\frac{1}{2}$ guineas, 0·73 of 16s. 8d., and ·375 of a florin.

27. There are twenty books on a shelf. Each book is $2\frac{5}{16}$ in. thick and there is a space of $\frac{1}{32}$ in. between each two books. How many feet and inches of the shelf are occupied?

28. An alloy is made of 36 parts copper, 2 parts brass, and 7 parts tin. What weight of tin is there in an alloy weighing 300 lb.?

29. Express £2. 15s. 6d. wholly in £'s.

30. Express 5·325 chains in yards, feet and inches.

CHAPTER II

RATIO AND PROPORTION

30. It has already been shown that a fraction may have its numerator and denominator multiplied or divided by the same number without having its value altered. For example $\frac{2}{8}=\frac{1}{4}=\frac{4}{16}$. The *relation* of each numerator to its respective denominator remains the same. The word *relation* just used is shortened to the word **ratio**. A fraction may therefore be looked upon as the ratio of two numbers.

When two or more fractions or ratios are equal to one another a proportion is formed. Thus $\frac{5}{10}$ and $\frac{6}{12}$, being equal fractions, form a proportion.

A proportion is the equality of ratios.

Consider the following proportions:

$$\frac{4}{8}=\frac{6}{12}, \qquad \frac{3}{9}=\frac{4}{12}, \qquad \frac{3}{4}=\frac{15}{20}.$$

It will be noticed that in every case the product of the first numerator and second denominator is equal to the product of the second numerator and first denominator, viz.

$$4 \times 12 = 6 \times 8.$$
$$3 \times 12 = 4 \times 9.$$
$$3 \times 20 = 15 \times 4.$$

By this means we can test whether or not two fractions or ratios are equal to each other. It is called the method of **cross multiplication**.

Summarising, if $\dfrac{a}{b}=\dfrac{c}{d}$, then $ad=bc$.

31. When dealing with fractions as ratios, the numerator and denominator are called **terms**.

When three terms of a proportion are known, the fourth can always be found.

Example 1.
$$\frac{5}{8}=\frac{35}{N}.$$
$$5 \times N = 8 \times 35.$$
$$\therefore \quad N = \frac{8 \times 35}{5}$$
$$= 56.$$

EXAMPLE 2. What number bears the same ratio to 45 as 17 bears to 51?

$$\frac{N}{45}=\frac{17}{51}.$$
$$N \times 51 = 45 \times 17.$$
$$\therefore \quad N = \frac{45 \times 17}{51}$$
$$= 15.$$

EXAMPLE 3. If 14 books cost £1. 11s. 6d., find the cost of 8 books.

Here it may be said that 14 books bear the same ratio to their cost as 8 books bear to their cost. The following proportion is formed:

$$\frac{14 \text{ books}}{\text{£1. } 11s. \ 6d.} = \frac{8 \text{ books}}{\text{cost of 8 books}}.$$
$$\therefore \quad \text{£1. } 11s. \ 6d. \times 8 = 14 \times \text{cost}.$$
$$\therefore \quad \frac{\text{£1. } 11s. \ 6d. \times 8}{14} = \text{cost of 8 books}$$
$$= \text{£0. } 18s. \ 0d.$$

or it may be said that 14 books bear the same ratio to 8 books as the cost of 14 books bear to the cost of 8 books. The following proportion is formed:

$$\frac{14}{8} = \frac{\text{£1. } 11s. \ 6d.}{\text{cost of 8 books}}.$$
$$\therefore \quad 14 \times \text{cost of 8 books} = \text{£1. } 11s. \ 6d. \times 8,$$
$$\therefore \quad \text{cost of 8 books} = \frac{\text{£1. } 11s. \ 6d. \times 8}{14}$$
$$= \text{£0. } 18s. \ 0d.$$

Exercises 16

1. Find a number in the same ratio to 27 as 50 is to 15.

2. Complete the following proportion:

$$\frac{15}{24} = \frac{N}{16}.$$

3. What weight bears the same ratio to 21 lb. as £2. 6s. 8d. does to £49?

4. If 25 pistons are bought for £3. 2s. 6d., how many may be bought for £13. 7s. 6d.?

5. The cost of travelling 15 miles is £0. 1s. $10\frac{1}{2}d.$ Find the cost of travelling 136 miles.

6. If $5\frac{1}{2}$ yd. of belt cost £0. 8s. 3d., find the price of $3\frac{3}{4}$ yd.

7. If 72 bolts cost £0. 5s. 0d., how many may be bought for £0. 2s. 1d. ?

8. If 29 men earn £32. 5s. 3d. in a week, what will 17 men earn in the same time ?

9. If $\frac{5}{8}$ of a cargo is valued at £1275, what part of the cargo will be worth £714 ?

10. Find the price of 4 ft. of material which costs £0. 19s. $10\frac{1}{2}d.$ for $13\frac{1}{4}$ yd.

11. If $4\frac{5}{8}$ shares are worth £33. 18s. 4d., find the value of 6 shares.

12. A house which cost £875 to build was let at a rental of £0. 16s. 8d. per week. What did it cost to build a house, the rental of which was £1 per week ?

13. If $\frac{3}{5}$ of a share is worth £46. 14s. 6d., find the value of $5\frac{1}{3}$ shares.

14. Gas costs £0. 4s. 2d. per 1000 cu. ft. How much has been consumed when the bill amounts to £1. 2s. 3d. ?

15. If 3 cwt. 28 lb. of brass sold for £12. 5s. 0d., what are 5 cwt. 56 lb. worth? (Answer to nearest penny.)

16. If £0. 9s. $7\frac{1}{2}d.$ is paid for $3\frac{1}{2}$ dozen rivets, find the cost of 27.

17. Eight planes are worth 21 chisels, and 70 chisels cost £8. 15s. 0d. Find the value of 20 planes.

18. A man bought 144 yd. of wire for £5. 8s. 0d. He used 252 ft., what was the value of the remainder?

19. The charge for travelling 27 miles is £0. 2s. 6d. How far can one travel for £4. 15s. 0d. ?

20. A wire rope 1·6 in. in diameter weighs 8 lb. per ft. length. Find the weight of 3 yd. 2 ft. of a rope the diameter of which is ·55 in.

INVERSE PROPORTION

32. The ratios and proportionals dealt with in the foregoing are *direct ratios and proportionals*, because as the value of a quantity increased or decreased in a certain ratio so the value of the dependent quantity increased or decreased in like ratio. Thus

$$\frac{27}{3} = \frac{36}{4}.$$

Then $\qquad \frac{81}{3} = \frac{108}{4}.$

But if one quantity depends on another quantity such that as the first quantity increases so the second quantity decreases, or *vice versa*, these quantities are said to be *inversely proportional*.

Thus a man walking at 2 m.p.h. does a certain journey in 6 hours. It will only take him half the time to do the journey if his rate were 4 m.p.h. When the speed is doubled the time is halved. The time is *inversely proportional* to the speed or rate.

EXAMPLE 1. A train travelling at 40 m.p.h. completes a journey in 8 hours. How long would the journey have taken had the rate been 30 m.p.h.?

As the rate has *decreased* so the time taken on the journey will *increase* in the same ratio.

$$\therefore \quad \text{Time} = 8 \times \tfrac{4}{3} = 10\tfrac{2}{3} \text{ hours.}$$

EXAMPLE 2. If 18 men complete a piece of work in 25 days, how long will it take 15 men to do an equal amount?

Here the number of men is *decreased*, therefore the time, 25 days, will be *increased proportionally*.

$$\therefore \quad \text{Time} = 25 \times \tfrac{18}{15} = 30 \text{ days.}$$

Exercises 17

1. If 16 men complete a piece of work in 18 days, how many men will do an equal amount in 24 days?

2. A room requires 108 yd. of paper 22 in. wide. How many yards will be required if the paper is 33 in. wide?

3. A wheel 2 ft. 3 in. in diameter revolves 2720 times in travelling a certain distance. How many revolutions would a wheel make whose diameter is 4 ft. 3 in.?

4. A train travelling at 35 m.p.h. can do a journey in 10 hours. At what rate will it have to travel, in order to do the journey in $8\tfrac{3}{4}$ hours?

5. In travelling a certain distance one wheel makes 1610 revolutions, while another makes 2070. If the diameter of the smaller wheel is 3 ft. 6 in., find the diameter of the larger wheel.

6. Under a pressure of 250 lb., the volume is 0·264 c.c.; what is the pressure when the volume is reduced to 0·044 c.c.?

7. In travelling a certain distance a wheel made 3553 revolutions. A new tyre was fixed, and in going the same distance only 2907 revolutions were made. If the diameter of the wheel at first was 4 ft. 6 in., what was the thickness of the new tyre?

8. The electrical resistance of a wire varies inversely as the square of its thickness. A wire $\frac{1}{8}$ in. in thickness has a resistance of 2·7 ohms. What resistance will a wire have whose thickness is $\frac{3}{16}$ in. ?

9. A drill whose diameter is $\frac{3}{4}$ in. makes 400 revolutions per minute. What is the diameter which makes 960 per minute, the peripheral speed in each case being the same?

10. The area of a pond is 3542 sq. ft., and its depth 5 ft. 6 in. By how much has its area been decreased if the depth of the water increases 11 in. ?

AVERAGES

33. If five rods A, B, C, D, E, 15, 13, 6, 16 and 5 in. long respectively (Fig. 9 A), are placed end to end, they would measure 55 in. Take another rod 55 in. long, and divide it into 5 equal parts P, Q, R, S, T. Place these equal lengths beside those of the first, as shown.

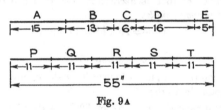

Fig. 9A

It is seen that these equal lengths are in some cases longer, and in others shorter than the unequal rods. But if the amount by which A, B and D are longer than the equal lengths is cut off, and added to the two rods C and E, there will just be enough to make these two rods each the same length as the others. This is called the **average length**.

To find the average value of a number of unequal values, add the unequal values together, and divide the sum by the number of values taken.

Example. A postman makes the following delivery of letters for a week: Monday 476, Tuesday 315, Wednesday 272, Thursday 374, Friday 202, Saturday 167, Sunday 0.

$$\text{Total delivered during week} = 1806.$$
$$\text{Number of days} \qquad = \quad 7.$$
$$\therefore \quad \text{Average daily delivery} = \frac{1806}{7}$$
$$= \quad 258.$$

Exercises 18

Find the average of the following:

1. 27, 42, 3, 18, 5.

2. 51, 36, 0, 47, 22, 6.

3. $3\frac{1}{2}$, $5\frac{1}{4}$, $3\frac{2}{3}$, $7\frac{1}{12}$.

4. Find the average daily cost of meals for a week. Monday 5s. 1d., Tuesday 5s. 8d., Wednesday 7s. 4d., Thursday 6s. 3d., Friday 4s. 2d., Saturday 6s. 7d., Sunday 5s. 9d.

5. A batsman makes the following scores in eight matches: 23, 72, 12, 156, 365, 41, 0, 83. What was his average?

6. Find the average cost of the following: 15 articles at £0. 4s. 4d. each, and 25 at £0. 3s. 6d. each.

7. The average of 15 results is 23. The average of the first 7 was 18·4 and of the last 7 it was 26·2, what was the 8th result?

8. The following kinds of metals are purchased: 48 lb. at £0. 2s. 8d. per lb., 16 lb. at £0. 2s. 6d. per lb., and 8 lb. at £0. 3s. 0d. per lb. What average price per lb. was paid?

9. The average weight of a football team including the goal keeper is 11 st. 3 lb., but excluding him the average is 11 st. What is the weight of the goal keeper?

10. A mixture is made up of 8 gall. worth £0. 1s. $10\frac{1}{4}d$. a gall., and 9 gall. worth £0. 1s. 6d. a gall. To this mixture is added 3 gall. of water. The mixture is sold at £0. 1s. 5d. per gall. What was the total gain or loss?

11. 150 yd. of wire cost £1. 6s. $1\frac{1}{2}d$. The average cost of 48 yd. was $1\frac{3}{4}d$. per yd. What was the average cost per yard of the remainder?

12. A pole 26 ft. 5 in. long was sawn into lengths of $11\frac{1}{4}$ in. If there were 28 of these lengths, how much wood was wasted per cut?

13. On a journey of 240 miles 6 gall. of petrol were used. On another journey of 190 miles 5 gall. were required, and on a third journey of 116 miles the car used 3 gall. Find the average number of miles to the gallon.

14. In testing a machine the following values of the efficiency were found: 89·76, 87·92, 91·04, 88·89, 89·43, 88·4. Find the average efficiency.

15. A man earned the following amounts in successive weeks: £3. 12s. 6d.; £4. 5s. 9d.; £3. 17s. 10d.; £5. 1s. 8d.; £4. 12s. 11d.; £6. 18s. 10d. What was his average weekly wage?

PERCENTAGES

34. A percentage is a fraction whose denominator is 100; that is, a percentage is so many parts of 100. Thus $\frac{5}{100}$ may be read 5 per cent.

Any fraction may be expressed as a percentage, by changing its denominator to 100 by multiplication or division.

EXAMPLE 1. Express $\frac{1}{4}$ as a p.c.

$$\frac{1}{4} = \frac{25}{100} = 25 \text{ p.c.}$$

EXAMPLE 2. Express $\frac{1}{3}$ as a p.c.

$$\frac{1}{3} = \frac{33\frac{1}{3}}{100} = 33\frac{1}{3} \text{ p.c.}$$

EXAMPLE 3. Express $\frac{5}{12}$ as a p.c.

To find what number 12 must be multiplied by to have 100 as the product, divide 100 by 12.

$$\frac{100}{12} = 8\frac{1}{3}.$$

$$\therefore \quad \frac{5}{12} = \frac{5 \times 8\frac{1}{3}}{12 \times 8\frac{1}{3}} = \frac{41\frac{2}{3}}{100} = 41\frac{2}{3} \text{ p.c.}$$

The same result may be obtained by multiplying $\frac{5}{12}$ by 100.

To express a fraction as a percentage multiply it by 100.

EXAMPLE 4. Express $\frac{2}{15}$ as a p.c.

$$\frac{2}{15} \text{ of } 100 = \frac{200}{15} = 13\frac{1}{3} \text{ p.c.}$$

EXAMPLE 5. Express $43\frac{3}{4}$ p.c. as a vulgar fraction.

$$43\frac{3}{4} \text{ p.c.} = \frac{43\frac{3}{4}}{100} = \frac{175}{400} = \frac{7}{16}.$$

EXAMPLE 6. Find the value of 5 p.c. of £105. 11s. 8d.

$$5 \text{ p.c.} = \frac{5}{100} = \frac{1}{20}.$$

$\frac{1}{20}$ of £105. 11s. 8d. $= 105$ shillings $+ 7d.$
$$= £5. \ 5s. \ 7d.$$

EXAMPLE 7. What percentage of £42 is £2. 2s. 0d.?

Express £2. 2s. 0d. as a fraction of £42.

$$\frac{\text{£2. 2s. 0d.}}{\text{£42}} = \frac{42}{42 \times 20} = \frac{1}{20}.$$

∴   percentage $= \frac{1}{20}$ of $100 = 5$ p.c.

Exercises 19

1. Express the following fractions as percentages: $\frac{1}{4}$; $\frac{3}{8}$; $\frac{7}{10}$; $\frac{5}{12}$; $\frac{2}{15}$; $\frac{5}{24}$; $\frac{7}{32}$; $\frac{8}{45}$.

2. Express the following percentages as fractions in their lowest terms: 16; $33\frac{1}{3}$; 45; $2\frac{1}{2}$; 75; 5; 28; $16\frac{1}{9}$.

3. Find the value of 5 p.c. of £1056. 11s. 8d.

4. $17\frac{1}{2}$ p.c. of 3 miles.

5. 16 p.c. of 25 gall.

6. 15 p.c. of 1 ton.

7. Express £1. 3s. 9d. as p.c. of £47. 10s. 0d.

8. Express 3 lb. 8 oz. as p.c. of 1 cwt.

9. If 30 p.c. of a certain sum is £14. 7s. 9d., what is the total amount?

10. What is the average of $12\frac{1}{2}$ p.c. of 144, 15 p.c. of 240, and $16\frac{2}{3}$ p.c. of 360?

11. A bar 3·15 ft. long on being heated expands to 3·43 ft. What is the percentage increase?

12. A cistern contains 250 gall. 4 p.c. of this is used, then 5 p.c. of the remainder leaks away. If now, 25 p.c. of what now remains is sold, how many gallons finally remain?

13. A person's income is £1050 per annum. Of this $7\frac{1}{2}$ p.c. is required for rent and rates. His insurances require 3 p.c. of the remainder. His spending amounts to £750. How much does he save?

14. A car was sold at a profit of 17 p.c. The selling price was £819. What was the cost price?

15. A casting which weighs 2 cwt. 2 qr. 20 lb., is composed of 93 p.c. copper, and the remainder tin. What weight of tin is in the alloy?

16. The result of an experiment gave that 1 cu. in. of cast iron weighed 0·28 lb. The result should have been 0·26 lb. per cu. in. Find the percentage error.

17. The expense of running a car was made up as follows: tyres £17, petrol £26, oil £2. 10s., tax and licence £11. 5s., repairs £3. 5s. 0d. Express each item as a percentage of the total cost.

18. A man whose salary was increased by £35 now gets £595. What percentage increase did he receive on his salary?

19. If $2\frac{1}{2}$ pints of water are added to $2\frac{1}{2}$ gall. of cutting fluid what percentage of the mixture is water?

20. A traveller receives $2\frac{1}{2}$ p.c. commission on his sales. What must his sales realise, in order that his commission may amount to £25. 17s. 6d. ?

CHAPTER III

POWERS AND ROOTS

35. The continuous product of two or more equal factors produces a number which is called a **power** of the factors.

Thus $2 \times 2 \times 2 = 8$, $5 \times 5 \times 5 \times 5 = 625$.

8 is called the third power of 2 because three factors 2 are used, and 625 is called the fourth power of 5 because four factors 5 are used. Instead of writing $2 \times 2 \times 2$ and $5 \times 5 \times 5 \times 5$, we write 2^3 and 5^4 and read 2 *to the power of* 3 and 5 *to the power of* 4 respectively. Conversely, 2 is called the 3rd (or cube) **root** of 8 and 5 is called the 4th **root** of 625.

36. Again, $7^2 = 49$ is read 7 *squared equals* 49. 7 is called the **square root** of 49. *The square root of a number is the number which when multiplied by itself will give the number.* Thus 4, 5, 8 are respectively the square roots of 16, 25 and 64. We write $\sqrt{16} = 4$, $\sqrt{25} = 5$, $\sqrt{64} = 8$.

37. The third, fourth and higher roots of a number are the numbers which when raised to the third, fourth or higher powers will produce the number. Thus 2 is the third root of 8 and 5 the fourth root of 625.

At present square roots only will be dealt with. Higher roots will be dealt with in a later chapter.

The sign $\sqrt{}$ is placed over a number whose square root is required.

Thus $\sqrt{100} = 10$.

$\sqrt{625} = 25$.

38. A number which has an exact square root is a **perfect square**. The square root of a perfect square can always be found by resolving it into its *prime factors*.

EXAMPLE 1. Find the square root of 1764.

$$1764 = 2 \times 2 \times 3 \times 3 \times 7 \times 7$$
$$= 2^2 \times 3^2 \times 7^2.$$
$$\therefore \quad \sqrt{1764} = 2 \times 3 \times 7 = 42.$$

The following method is more generally used:

EXAMPLE 2. Find the square root of 20736.

```
1          2'07'36        |144
1          1
‾‾         ‾‾
24         107
 4          96
‾‾‾        ‾‾‾‾
284        1136           ∴   √20736 = 144.
           1136
```

1. Mark off the number in periods of two, commencing at the decimal point.

2. Find the number whose square is nearest to (equal to or less than) 2. Here it is 1.

3. Place the figure 1 in the quotient and at the same time place a figure 1 to the left of the number.

4. Square 1 and place the result below the first period. Subtract, leaving 1 as remainder.

5. Bring down the next period 07 and place it after the remainder just found making 107, at the same time double the 1 on the left-hand side. This now forms the new trial divisor.

6. Divide 107 by the trial divisor 2. This divides into 10, five times, but when 5 is placed beside the 2 and the result multiplied by 5, it gives a number too great to be subtracted from 107. Therefore 4 is substituted for the 5. Multiplying by 4, gives a result 96, which is subtracted from 107, leaving a remainder of 11.

7. Bring down the next period 36 and place it after the 11, at the same time double 4 in the left-hand column. The trial divisor is now 28. 2 divides into 11 five times, but when the 5 is placed after the 8 and multiplied by 5 in the quotient, too large a number is produced to subtract from 1136. The next lower figure is then tested, placed after the 8. The number 284 is multiplied by 4 and produces 1136 exactly. Therefore 144 is the square root of 20736.

39. Square root of a decimal fraction.

The number is marked off in periods of two, commencing at the decimal point, and the same procedure is followed. If there is an odd number of figures in the decimal portion of the number a cipher is added.

40. Square root of a vulgar fraction.

This may be done in two ways:

1. The vulgar fraction may be expressed as an equivalent decimal fraction, and the square root then found.

2. If the denominator of the fraction is not a perfect square, make it so by multiplying it by itself, at the same time multiplying the numerator by the same number. The square root of the numerator may now be found, and if it is not a perfect square ciphers may be brought down after the decimal point so as to give an approximate square-root value for the numerator. The denominator is a perfect square, whose square root is divided into the approximate square root of the numerator for the final result.

EXAMPLE. Find the square root of $\frac{6}{7}$.

$$\sqrt{\frac{6}{7}} = \sqrt{\frac{6 \times 7}{7 \times 7}} = \frac{\sqrt{42}}{7}.$$

		6·48074
6	42	
6	36	
124	600	
4	496	
1288	10400	
8	10304	
129607	960000	
7	907249	
1296144	5275100	
	5184576	
	90524	

$$\therefore \quad \frac{\sqrt{42}}{7} = \frac{6·48074}{7} = 0·92582.$$

Exercises 20

Find the square roots of the following:

1. 5929.
2. 7056.
3. 5041.
4. 1156.
5. 8464.
6. 56.
7. 87.
8. 287.
9. 153·76.
10. 841·5801.
11. 3696·64.
12. ·0625.
13. $\frac{5}{6}$.
14. $\frac{3}{8}$.
15. $\frac{5}{9}$.
16. $1\frac{4}{5}$.
17. $2\frac{2}{7}$.
18. $5\frac{5}{16}$.
19. $7\frac{5}{13}$.
20. $9\frac{1}{9}$.
21. $3\frac{1}{7}$.

CHAPTER IV

SYMBOLS

41. A **symbol** is a letter or character used to represent some definite value, or to express a relation between two or more different quantities. Thus the statement that the area of a rectangle is found by multiplying its length by its breadth can be expressed by means of symbols. Putting **L** for length, **B** for breadth, and **A** for area, the statement becomes

$$LB = A.$$

It is known that the circumference of a circle is $3\frac{1}{7}$ times its diameter. This may be expressed

$$C = 3\tfrac{1}{7}D.$$

42. This method of dealing with symbols instead of with actual values is called **Algebra** and the above statements concerning the rectangle and the circle, expressed by the use of symbols, are called algebraic formulae or algebraic expressions.

43. It will be noticed in both formulae used above, that although multiplication is implied in **LB** and $3\frac{1}{7}$**D**, the sign of multiplication is not used. *The usual method adopted in algebra is to omit signs of multiplication and division.* To express division of one quantity by another the quantities are written in fractional form. Thus $\dfrac{M}{2}$ means that M is divided by 2. $\dfrac{PZ}{H}$ means that P and Z are multiplied and the result is divided by H. The symbols so used are called the **terms** of the expression.

44. Consider the expressions $3\frac{1}{7}D$, LB, PZ. In these $3\frac{1}{7}$, L and P are called the **coefficients** of D, B and Z respectively. $3\frac{1}{7}$ is a *numerical coefficient* and L and P are *literal coefficients*.

45. When definite values are given to symbols, so that a definite value may be arrived at for the whole expression, the process is called **evaluation by substitution**.

EXAMPLE 1. Find the value of $3a - bc + \dfrac{c}{2}$, when $a = 8$, $b = 6$, $c = 4$.

$$3a - bc + \frac{c}{2}$$
$$= 3 \times 8 - 6 \times 4 + \tfrac{4}{2}$$
$$= 24 - 24 + 2$$
$$= 2.$$

EXAMPLE 2. Evaluate $m^2 + 3yv^3 - \dfrac{x^2}{4}$, when $m = 3$, $y = 6$, $v = \tfrac{1}{2}$, $x = 5$.

$$m^2 + 3yv^3 - \frac{x^2}{4}$$
$$= 3 \times 3 + 3 \times 6 \times \tfrac{1}{2} \times \tfrac{1}{2} \times \tfrac{1}{2} - \frac{5 \times 5}{4}$$
$$= 9 + \tfrac{18}{8} - \tfrac{25}{4}$$
$$= 9 + 2\tfrac{1}{4} - 6\tfrac{1}{4}$$
$$= 5.$$

Exercises 21

Find the value of the following when $x = 4$, $y = 5$, $a = xy$, $b = 2x$.

1. $xy + 2x - 5y$.

2. $a^2b - \dfrac{3ax^2y^2}{10}$.

3. $\dfrac{xy - b}{2} + \dfrac{4y - a}{b} + 1$.

4. $2a - 2b + 19$.

5. $abx - 2x^3y$.

6. $S = \tfrac{1}{2}gt^2$. If $S = 25$, $g = 32$, find t.

7. $V^2 = 2fs$. If $f = 149$, $s = 72$, find V.

8. $T = 2\pi \sqrt{\dfrac{l}{g}}$. If $\pi = 3 \cdot 14$, $l = 50$, $g = 32$, find T.

9. $S = 2\pi r^2 + 2\pi rh$. If $\pi = 3 \cdot 14$, $h = 5 \cdot 9$, $r = 3 \cdot 1$, find S.

10. $S = \dfrac{A^2h}{3} + \dfrac{Aah}{3} + \dfrac{a^2h}{3}$. If $A = 3$, $a = 2 \cdot 2$, $h = 2 \cdot 5$, find S.

ALGEBRAIC SUMS

46. The signs $+$ and $-$ have opposite meanings; one is to *increase* the other to *decrease*; or if one represents *movement to the right*, then the other will represent *movement to the left*. The $+$ sign usually represents movement to the right, or positive (written $+$ ve) direction; and $-$ represents movement to the left, or negative (written $-$ ve) direction. The same convention is applied to upward and downward movements, upward being $+$ ve and downward $-$ ve direction.

EXAMPLE 1. A sliding piece moves 25 in. towards the right, and then 16 in. towards the left. How far now is it from its starting position?

$+25$ represents its first movement and -16 represents its second movement.

Therefore $+25-16$ represents both movements the result of which is $+9$, that is, it is now 9 in. to the right of its starting point. Thus the algebraic sum of $+25$ and -16 is $+9$.

EXAMPLE 2. A thermometer rises through 8 degrees, and then falls through 14 degrees. What is now the position of the top of the mercury column?

The first movement is $+8$, the second movement is -14; that is, it continues falling until it reaches its starting point and then drops for 6 more degrees. Finally, therefore, it is 6 below the starting point and this position is represented by -6.

$$\therefore \quad 8-14=-6.$$

-6 is the algebraic sum of $+8$ and -14.

It will be noticed that in Examples 1 and 2 *the algebraic sum bears the sign of the greater quantity*. This is always the case.

Rule. *If several quantities are connected by the signs $+$ and $-$, add together quantities of the same kind having a $+^{ve}$ sign for the total $+^{ve}$ result. Then add like quantities having a $-^{ve}$ sign for the $-^{ve}$ result. The algebraic sum of these two is the final result.*

EXAMPLE 3. Find the algebraic sum of $5a-3a+a+4a-2a$
$$= 5a+a+4a-3a-2a$$
$$= 10a-5a$$
$$= 5a.$$

(Steps: (1) *add the positives*, (2) *add the negatives*, (3) *find the difference*, (4) *affix the sign of the greater*.)

EXAMPLE 4. Simplify $4a-3c-2a+6c-2c+a$
$$= 4a+a-2a+6c-2c-3c$$
$$= 5a-2a+6c-5c$$
$$= 3a+c.$$

Exercises 22

Simplify the following:

1. $x-3y+4x-2y-8x+7y.$ 2. $a-b+2c+b+c-3a.$

3. $3x^2+2x-8+5x-5x^2+4+4x^2-9x.$

4. $4 - 2x + 2 - 2x^2 + 6 - 4x + 3x^2.$
5. $3x^2 - 2x - 2 - 2x^2 + 6x.$
6. $4p^3 + 2p^2 - 3 + 4p^2 - 3p^3 + 8 - 7p^2.$
7. $\frac{3}{4}a^2 + \frac{1}{2}a + 4 + \frac{3}{2}a^2 + \frac{3}{4}a - 8 - \frac{1}{2}a^2.$
8. $p^2 - 3q + 2p + 4q - 4p^2 - 3q + 5.$
9. $2x^3 - 2x^2 + x + 3x^3 + 5x^2 + 4x - 4x^3 - 3x^2 - 3x.$
10. $3a^2 - \frac{1}{5}ab + \frac{1}{2}b^2 - \frac{5}{2}a^2 + ab - \frac{3}{2}b^2 - \frac{1}{2}a^2 + 2ab + b^2.$

47. Algebraic multiplication and division.

The meaning of the expression 2^3 has already been explained (para. 35).

If 2^3 is multiplied by 2^4 the result will be $2 \times 2 \times 2 \times 2 \times 2 \times 2 \times 2 = 2^7$, that is 2^{3+4}. In the same way if x is substituted for 2, $x^3 \times x^4 = x^7$.

To multiply two or more quantities of the same kind together, the sum of the powers of the factors will give the power of the product.

EXAMPLE. $\qquad x^2 \times x^3 \times x = x^{2+3+1} = x^6.$

If the quantities multiplied together have coefficients, these are multiplied together in the ordinary way. Thus:

$$3a^2 \times 4a^3 = 12a^5.$$
$$ab^2 \times 2b = 2ab^3.$$

Powers of *different* quantities multiplied together cannot be added. Thus

$$p^2 \times m^3 = p^2 m^3.$$

48. In division of quantities of the same kind by one another, the power of the result is the difference between the powers of the dividend and divisor.

Thus $\qquad x^5 \div x^3 = \dfrac{\cancel{x} \times \cancel{x} \times \cancel{x} \times x \times x}{\cancel{x} \times \cancel{x} \times \cancel{x}} = x \times x = x^2 = x^{5-3}.$

The coefficient of the dividend is divided by the coefficient of the divisor in the ordinary way. Thus

$$15a^4 \div 3a = 5a^3.$$

49. The meaning of x^{-3}.

It has been shown in para. 46 that in an algebraic sum, the result always bears the sign of the greater quantity, thus

$$9 - 6 = 3, \quad 3 - 12 = -9.$$

If therefore a^2 is divided by a^5, the result is $a^{2-5}=a^{-3}$. Again this division may be carried out in another way, thus

$$a^2 \div a^5 = \frac{\not{a} \times \not{a}}{\not{a} \times \not{a} \times a \times a \times a} = \frac{1}{a^3}.$$

$$\therefore \quad a^{-3} = \frac{1}{a^3}.$$

In general $a^{-p} = \dfrac{1}{a^p}$.

50. Rule of signs.

When two or more quantities, the signs of which are alike, are multiplied together the product is positive; if the signs are different the sign of the product is negative.

Like signs produce plus.
Unlike   ,,   ,,   minus.

Thus
$$2x^2 \times -3x^3 = -\ 6x^5.$$
$$4p^3 \times\ \ 2p =\ \ \ 8p^4.$$
$$-3a\ \times -5a^3 =\ \ 15a^4.$$

The same rule applies to division.

Thus
$$12a^4 \div -3a\ = -4a^3.$$
$$16p^5 \div\ \ 8p^4 =\ \ 2p.$$
$$-21b^4 \div -7b^2 =\ \ 3b^2.$$

EXAMPLE 1. Multiply $x^3 - 2x^2 + 1$ by x^4. Each term is multiplied separately by x^4. Thus $x^3 - 2x^2 + 1 \times x^4 = x^7 - 2x^6 + x^4$.

EXAMPLE 2. Multiply $2a^2 + 3a$ by $3a + 2$.

$$
\begin{array}{l}
\quad\quad\quad 2a^2 + 3a \\
\quad\quad\quad 3a\ + 2 \\
\hline
6a^3 +\ 9a^2 \\
\quad\quad +\ 4a^2 + 6a \\
\hline
6a^3 + 13a^2 + 6a.
\end{array}
$$

Multiply by $3a$,   ,,   2

EXAMPLE 3. Multiply $4y - c$ by $3y^2 - c^2$.

$$
\begin{array}{l}
\quad\quad 4y\ - c \\
\quad\quad 3y^2 - c^2 \\
\hline
12y^3 - 3y^2c \\
\quad\quad\quad\quad - 4yc^2 + c^3 \\
\hline
12y^3 - 3y^2c - 4yc^2 + c^3.
\end{array}
$$

Multiply by $3y^2$,   ,, $-c^2$

EXAMPLE 4. Divide $4x^3 - 6x^2 + 8x$ by $2x$.

Each term is divided separately.

$$4x^3 - 6x^2 + 8x \div 2x = 2x^2 - 3x + 4.$$

EXAMPLE 5. Divide $12p^5 - 4p^3 + 4p^2$ by $-4p^2$.

$$12p^5 - 4p^3 + 4p^2 \div -4p^2 = -3p^3 + p - 1.$$

Divisor containing two or more terms.

1. *See that Divisor and Dividend are arranged in ascending or descending powers of some common quantity.*

2. *When so arranged, divide the left-hand quantity in the divisor into the left-hand quantity in the dividend. The result is the first term of the quotient.*

3. *Now multiply the whole divisor by this term in the quotient, and place the products under corresponding terms of the dividend. If there are no corresponding terms set them out by themselves.*

4. *Next subtract the product so obtained in no. 3 above from the dividend. Next bring down as many terms as necessary from the dividend and place them beside the remainder just obtained by subtraction.*

5. *Repeat the foregoing process, until all the terms of the dividend are exhausted.*

EXAMPLE. Divisor)   Dividend   (Quotient

$\qquad\qquad$ $x - 3$) $2x^2 + 4x - 30$ (

Divide x into $2x^2$. The result is $2x$. Place $2x$ in the quotient and multiply divisor by $2x$, putting the product below the corresponding terms of the dividend. Thus:

$$x - 3 \ ) \ 2x^2 + 4x - 30 \ (\ 2x$$
$$\underline{2x^2 - 6x}$$

Subtract $2x^2 - 6x$ from $2x^2 + 4x$. The result is $10x$. Bring down -30 beside $10x$ and the solution now stands:

$$x - 3 \ ) \ 2x^2 + 4x - 30 \ (\ 2x$$
$$\underline{2x^2 - 6x}$$
$$10x - 30$$

Divide x into $10x$. This gives 10. Place 10 in the quotient, and multiply divisor by 10. Place the product below the dividend and subtract. This gives

$$x - 3 \,)\, 2x^2 + 4x - 30 \,(\, 2x + 10$$
$$\underline{2x^2 - 6x}$$
$$10x - 30$$
$$\underline{10x - 30}$$

The result is $2x + 10$.

Exercises 23

1. $5ax \times 3a^2x.$ **2.** $3xy^2 \times -2x^2y.$

3. $3x^2 - y^2$ multiply by $2xy^3.$

4. $x^2 + 9$,, $x^2 - 8.$

5. $a^2 - ax + x^2$,, $a + x.$

6. $4a - 5b$,, $3a - 2b.$

7. $x^2 + xy + y^2$,, $x - y.$

8. $x^2 - 2x + 1$,, $x + 2.$

9. $2x^3 - 4x - 3$,, $2x + 3.$

10. $\frac{1}{2}a^2 + \frac{1}{3}a - 4$,, $\frac{1}{2}a - 3.$

11. $ab^2 \div ab.$

12. $-15x^3y^2 \div 3xy.$

13. $-42x^2y^3z^2 \div -6x^2yz.$

14. Divide $2a^3 + 5a^2 - 2a - 15$ by $2a - 3.$

15.   ,, $12x^2 - 16 + 3x^3$,, $3x + 4.$

16.   ,, $45 - 21x^2 + 9x - 12x^3$,, $5 - 4x.$

17.   ,, $x^4 - x^3 - 8$,, $x - 2$

18.   ,, $x^2 - 4x^3 - 9 + 4x^4$,, $x + 1.$

19.   ,, $2x^3 - x^2 - 8x + 4$,, $2x - 1.$

20.   ,, $a^3 - 6a^2b + 12ab^2 - 8b^3$,, $a - 2b.$

51. Brackets and symbolic expression.

If two or more terms are to be regarded as a single quantity, they are enclosed in a bracket. Thus in the expression $a - (b + c)$, the sum of b and c is to be subtracted from a. Or in the expression $p + 2\,(r - s)$, twice the difference between r and s is to be added to p.

EXAMPLE 1. Evaluate $8-(3+2)$.

Here the sum of 3 and 2 is to be subtracted from 8.

$$\therefore \quad 8-(3+2)$$
$$=8-5$$
$$=3.$$

But the same result would be obtained if 3 were first subtracted from 8, and then 2 from the answer obtained, that is, $8-(3+2)$ might be written $8-3-2$. From this the following rule may be inferred:

When a minus sign is in front of a bracket, the signs inside are changed, when the bracket is removed.

EXAMPLE 2. Simplify $a+2b-(2a-3b+c)$.

$$a+2b-(2a-3b+c)=a+2b-2a+3b-c$$
$$=-a+5b-c.$$

52. The converse of the above rule also must be observed:

When a number of terms are to be enclosed in a bracket, the signs of all terms, so enclosed, are changed when the first term included is a minus quantity, but no change takes place if the first term is a positive quantity.

EXAMPLE 3. In the expression $a+2b-x-2y+z$, enclose the last three terms in a bracket.

$$a+2b-x-2y+z=a+2b-(x+2y-z).$$

The expression $15-2(5+2)$ means that twice the sum of 5 and 2 is to be subtracted from 15, that is, the expression becomes $15-2\times7=15-14=1$. The same result would be obtained by first removing the bracket, and at the same time multiplying each term within the bracket separately, observing the rule of signs, namely

$$15-2(5+2)=15-10-4$$
$$=15-14=1.$$

If a quantity is placed immediately in front of a bracket, all terms inside the bracket are multiplied by that quantity, when brackets are removed.

53. If one pair of brackets is enclosed within another pair, the inner pair should be removed first.

EXAMPLE 4. Simplify $3x - 4\{3x - 5(4x - 6)\} - 35$
$$= 3x - 4\{3x - 20x + 30\} - 35$$
$$= 3x - 12x + 80x - 120 - 35$$
$$= 71x - 155.$$

Exercises 24

Remove the brackets from the following:

1. $a + 2b - (3a + b) + 3(a - 2b)$.

2. $x + 3(x - 2y) - (x + 5y - 2)$.

3. $2a - \{3b + 4c - (a - 2b + 4c)\}$.

4. $2x + 2\{y - (x - y) - 2y\} + x$.

Evaluate the following: $x = 5$, $y = 4$, $z = 3$.

5. $x - (y - z) - \{2x - 4y - 2z + 3(y + z)\}$.

6. $3x - 2\{2x - (3x - 3y + x) - y\}$.

7. $7x + 3(y - 3z) - 3\{x + 2(y - z)\} + 2$.

8. $[3x + \{4y - 2x - (3y + x) + y\}]$.

9. $[2a - \{3b + 4c - (a - 2b + 4c)\}]$.

10. $[2a - \{3b + 2(a - 2b) + b\}]$.

11. Enclose the last three terms in brackets:
$$x - 3y - 4x - 8y + 12.$$

12. Re-write the following considering the 2nd and 3rd terms as one quantity, and the 4th and 5th terms as another quantity:
$$a - 2b + c + 6a - 9b.$$

13. Write down the following in algebraic form. Subtract the sum of $3x$ and y from the difference between z and 4.

14. A man walks 28 miles in 3 days. He walks x miles the first day, and $2y$ miles the second day. Express algebraically how far he walks on the third day.

15. The length of a rectangle is f ft. and breadth m ft. Express its perimeter in inches.

16. From 5 times the excess of m over n subtract the product of x and y. Multiply the result by 2.

17. *A* walks at *w* miles per hour, *B* at (*m* + 2) miles per hour. If they start together, how far ahead will *B* be in 45 min.?

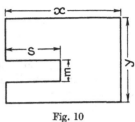

Fig. 10

18. A man has £*x*. He spends five shillings and then receives three times the money he has remaining less £2. How much has he finally?

19. A man walks at the rate of (*m* + 2) miles per hour. He is 25 miles from his destination. How far will he be from it 3 hours later?

20. What is the perimeter of this irregular octagon? (Fig. 10.)

CHAPTER V

FACTORS

54. The definition of a factor in Chapter I equally applies to an algebraic expression. Thus in the expression $a^2 + ac$, a is a factor of a^2 and also of ac, therefore $a^2 + ac$ may be written $a(a+c)$. This is the simplest form of an algebraic factor.

EXAMPLE 1. Factorise $a^2x + 2ay + 3ab$.

On examination, it is seen that a is common to all the terms.

$$\therefore \quad a^2x + 2ay + 3ab = a(ax + 2y + 3b).$$

EXAMPLE 2. Factorise $2x^3y + 4x^2y^2 + 6xy^3$.

In this example 2 is common to all the terms, so also are x and y; therefore $2xy$ is the common factor.

$$\therefore \quad 2x^3y + 4x^2y^2 + 6xy^3 = 2xy(x^2 + 2xy + 3y^2).$$

55. An expression of four terms may sometimes be factorised, by arranging the expression in two pairs of terms.

EXAMPLE 3. Factorise $ac + bd + ad + bc$.

This may be written in two pairs, each pair having a common factor.

Thus $\qquad ac + bd + ad + bc = (ac + ad) + (bc + bd)$
$$= a(c+d) + b(c+d).$$

Now $(c+d)$ is a common factor:

$$\therefore \quad a(c+d) + b(c+d)$$
$$= (c+d)(a+b),$$

or $\qquad ac + bd + ad + bc = (ac + bc) + (bd + ad)$
$$= c(a+b) + d(a+b)$$
$$= (a+b)(c+d).$$

56. Consider the products of the following pairs:

$$(x+4) \text{ and } (x+3).$$
$$(x+5) \quad,, \quad (x-3).$$
$$(x-2) \quad,, \quad (x-4).$$
$$(x+1) \quad,, \quad (x-3).$$

$$\text{Factors} \qquad \text{Product}$$
$$(x+4)\,(x+3)=x^2+7x+12.$$
$$(x+5)\,(x-3)=x^2+2x-15.$$
$$(x-2)\,(x-4)=x^2-6x+8.$$
$$(x+1)\,(x-3)=x^2-2x-3.$$

Examining the four products the following is noticed. *In each case*:

1. The *first term* is the product of the two first terms of the factors.

2. The *second term* is the algebraic sum of the two following products:

(a) The 2nd term of the first factor and the 1st term of the second factor (First Product).

(b) The 1st term of the first factor and the 2nd term of the second factor (Second Product).

3. The *third term* is the product of the two 2nd terms of the factors.

EXAMPLE 4. Factorise $x^2+8x+15$.

Clearly the first term of each factor is x. The second terms of the factors must be such that their algebraic sum is $+8$, and their product $+15$. Clearly these two terms must be $+3$ and $+5$.

$$\therefore \quad x^2+8x+15=(x+3)\,(x+5).$$

EXAMPLE 5. Factorise x^2-3x-4.

Here again the first terms of the factors must be x. The second terms of the factors will be two numbers whose algebraic sum is -3, and whose product is -4. The only numbers satisfying this condition are $+1$ and -4.

$$\therefore \quad x^2-3x-4=(x+1)\,(x-4).$$

EXAMPLE 6. Factorise $4x^2+8x+3$.

Here the two first terms may be either $4x$ and x, or $2x$ and $2x$. The two second terms are either 1 and 3, or -1 and -3, because the sign of the third term is positive.

Note. *If the sign of the third term is positive, both second terms of the factors are alike in sign, namely negative if the second term is negative, and positive if the second term is positive.*

In the above expression, as the second term $(+8x)$ is positive, both second terms of factors are positive.

Trial must be made of the various combinations that may be made of $4x$ and x, or $2x$ and $2x$, with 1 and 3, so as to produce $4x^2 + 8x + 3$. It is found that the factors are $(2x+1)(2x+3)$.

57. If in an expression consisting of three terms, the second term is twice the product of the square roots of the first and third terms, then the expression is called a **perfect square**. It is the square of the sum or difference (according as the sign of the second term is $+$ or $-$) of the square roots of the first and third terms. Thus in the expression $a^2 + 2ab + b^2$ the second term $(2ab)$ is twice the product of the square roots of a^2 and b^2.

$$\therefore \quad a^2 + 2ab + b^2 = (a+b)^2.$$

EXAMPLE 1. Factorise $\quad a^2 + 8a + 16$.

Here $\qquad\qquad 8a = 2\sqrt{a^2} \times \sqrt{16}.$

$$\therefore \quad a^2 + 8a + 16 = (a+4)^2.$$

EXAMPLE 2. Factorise $\quad x^2 - 4cx + 4c^2$.

Here $\qquad\qquad 4cx = 2 \times \sqrt{x^2} \times \sqrt{4c^2}.$

$$\therefore \quad x^2 - 4cx + 4c^2 = (x-2c)^2.$$

58. *The difference of the squares of two quantities is equal to the product of the sum of the square roots of the quantities and the difference of the square roots.*

Thus in $x^2 - y^2$, $\sqrt{x^2} = x$, $\sqrt{y^2} = y$

sum of square roots $= x + y$.

diff.   ,,   ,, $= x - y$.

$$\therefore \quad x^2 - y^2 = (x+y)(x-y).$$

Similarly $\qquad\qquad a^2 - b^2 = (a+b)(a-b).$

Similarly $\qquad\qquad \pi R^2 - \pi r^2 = \pi(R+r)(R-r).$

EXAMPLE. The end plate of a boiler is circular and has a diameter of D; four circular holes, each of diameter d, are cut in it for furnace pipes. Write down a formula for the remaining area of the plate and factorise it so as to put it in a form most suitable for calculation. Using your simplified formula, find the remaining area in square feet to three significant figures, when $D = 9$ ft., $d = 0.5$ ft. $(N.C.)$

$$\frac{\pi}{4} D^2 - 4 \times \frac{\pi}{4} d^2 = \text{area of remainder (para. 112)}$$

$$= \frac{\pi}{4} (D^2 - 4d^2)$$

$$= \frac{\pi}{4} (D + 2d) (D - 2d)$$

$$= \frac{\pi}{4} (9 + 2 \times 0.5) (9 - 2 \times 0.5)$$

$$= \frac{\pi}{4} (10) (8) = \frac{\pi}{4} \times 80$$

$$= 20\pi = 62.8 \text{ sq. ft.}$$

Exercises 25

Factorise the following:

1. $x^2 + 2px - 4x$.
2. $3p + 6p^2 - 15p^3$.
3. $\dfrac{x^2}{y^2} - \dfrac{2x^3}{y^3} + \dfrac{3x^4}{y^4}$:
4. $a^2 + bc + ac + ab$.
5. $pr - ps - qs + qr$.
6. $3ax - 2y + ay - 6x$.
7. $x^2 + 10x + 16$.
8. $x^2 - 2x - 15$.
9. $x^2 + x - 42$.
10. $x^2 + 2x - 35$.
11. $3a^2 - 3a - 60$.
12. $x^2 + 6ax - 72a^2$.
13. $2a^2 - 2a - 4$.
14. $a^2b^2 - 2ab - 48$.
15. $a^2 - 10a + 25$.
16. $4a^2 + 12a + 9$.
17. $9p^2 - 6ps + s^2$.
18. $36x^2 - 25y^2$.
19. Evaluate by factors 102×98.
20. Evaluate by factors 996×1004.

Exercises 26

More difficult exercises, in factors:

1. $2x^2 + 5x + 2$.
2. $3x^2 + 7x + 2$.
3. $6x^2 - 5x + 1$.
4. $2x^2 + 3x - 2$.
5. $17x^2 + 201x - 36$.
6. $6x^2 - 11x - 10$.
7. $6x^2 - 11x - 35$.
8. $8x^2 + 2x - 21$.

9. $4x^2 + 41x - 84.$

10. $15x^2 - 19x - 8.$

11. $12x^2 + 5x - 2.$

12. $x^2 - 4xy - 4K^2 + 4y^2.$

13. $a^2 + b^2 - c^2 + 2ab.$

14. $a^2 + b^2 - c^2 - 2ab.$

15. $(3x + 1)^2 - (2x - 1)^2.$

16. $1 - 2ab - a^2 - b^2.$

17. $4a^2 - x^2 - y^2 + 2xy.$

18. $4ax - 4ay - x^2 + y^2.$

19. $9x^4 - a^4 - b^4 + 2a^2b^2.$

20. $4 - x^2 + 4xy - 4y^2.$

CHAPTER VI

59. An equation is an algebraic statement that two expressions are equal.

Thus $2x = 8$ is an equation; $2x$ is called the *left-hand side of the equation* and 8 is called the *right-hand side*.

Equations are used for finding the value of some unknown quantity, which is denoted by a letter. Finding the value of the unknown quantity is termed **solving the equation**, and the value, so found for the unknown quantity, is said to **satisfy the equation**.

60. A simple equation is one which contains one unknown quantity only, and also that quantity expressed to the power of 1. Thus $2x = 8$ is a simple equation, but $2x^2 = 8$ is not a simple equation because the unknown quantity is expressed in its second power.

61. An equation may be compared with a physical balance, the equilibrium of which can only be preserved if equal additions or subtractions take place in both pans at the same time. As multiplication and division are shortened forms of addition and subtraction, the following statement holds good.

If both sides of an equation are increased or decreased by equal amounts the truth of the equation is unaltered.

Thus if $\qquad\qquad 5x = 10.$

Then $\qquad\qquad 5x + 3 = 10 + 3.$

$\qquad\qquad\qquad 5x - 3 = 10 - 3.$

$\qquad\qquad\qquad 5x \times 2 = 10 \times 2.$

$\qquad\qquad\qquad 5x \div 2 = 10 \div 2.$

62. Solving an equation.

EXAMPLE 1. Solve $\qquad 4x = 12.$

The value of x is required. x is one quarter of $4x$. Therefore its value is one quarter of 12.

$$\therefore \quad x = 3.$$

In other words, divide both sides of the equation by 4, i.e. by the *coefficient of the unknown*.

EXAMPLE 2. Solve $\qquad 3x + 4 = 13.$

When one side of an equation consists partly of the unknown quantity and partly of a number, the first step is to arrange the equation with only the unknown quantity, on one side (usually the left), and the numbers on the other. This is called **transposing the terms**. *When a term is transposed its sign is changed.*

$$3x + 4 = 13.$$
$$\therefore \quad 3x \quad = 13 - 4 \text{ (transpose 4 and change its sign)}.$$
$$\therefore \quad 3x \quad = 9.$$
$$\therefore \quad x \quad = 3.$$

EXAMPLE 3. Solve $\qquad 6x - 18 = 2x + 26.$

Transposing $\qquad 6x - 2x = 26 + 18.$
$$\therefore \qquad 4x = 44.$$
$$\therefore \qquad x = 11.$$

Exercises 27

Solve the following equations:

1. $3x = 24.$
2. $4x + 2 = 18.$
3. $7x = 2x + 15.$
4. $42x - 30 = 4 + 25x.$
5. $x - 3 + 2x - 2 = x - 2.$
6. $3x - 17 - 5x = 12 - 4x - 9.$
7. $30x - 500 = 340 - 40x.$
8. $4 + x - 1 = 8x - 2(3x - 4) + 3.$
9. $4x - 2(x + 4) = 3x - 13.$
10. $3x - 15 + 2x + 2 = x + 15.$
11. $4(x - 1) - 2(2x - 3) = 2x - 3.$
12. $7 - 3x + 3(x - 1) = 2x.$
13. $2(x - 3) - 3(x + 2) + 18 = 0.$
14. $3x - 2\{x + 2(x - 1)\} = 5x - 28.$
15. $6x - \{2x + 3(x - 3)\} = 5x - 4.$
16. $3x + 4(x - 1) = 5x - 2\{x + 2(x - 1)\}.$
17. $4x + 2(x - 2) - 3(x - 3) = 2(2x - 3) + x + 1.$
18. $2(x + 1) + 3(x + 2) - 4(x + 4) = 6(x - 3).$
19. $5x - 2(x + 3) - (x - 5) = 3(2x - 3) - 4.$
20. $5 - 3(2x - 1) + 8(x - 2) = 0.$

63. Fractional equations.

Equations involving fractions are treated in the same way as fractions in arithmetic. Find the common denominator and multiply both sides of the equations by this number.

EXAMPLE 1. Solve the equation $\dfrac{x}{2} + \dfrac{x}{3} - \dfrac{x}{4} + \dfrac{x}{5} = \dfrac{47}{6}$.

The common denominator is 60.

Multiplying every term by 60 the following is obtained:

$$30x + 20x - 15x + 12x = 470.$$
$$62x - 15x = 470.$$
$$\therefore \quad 47x = 470.$$
$$\therefore \quad x = 10.$$

EXAMPLE 2. Solve $\dfrac{x-5}{3} - \dfrac{x+7}{18} = \dfrac{x+1}{12}$.

Here the C.D. is 36.

Multiplying each term by 36 the following is the result:

$$12\,(x-5) - 2\,(x+7) = 3\,(x+1).$$
$$12x - 60 - 2x - 14 = 3x + 3.$$
$$12x - 2x - 3x = 3 + 14 + 60.$$
$$\therefore \quad 7x = 77.$$
$$\therefore \quad x = 11.$$

Exercises 28

Solve the following equations:

1. $\dfrac{4x}{5} - 27 = \dfrac{3x}{4} - 23$.

2. $6 + \dfrac{x}{5} = 7 + \dfrac{x}{6}$.

3. $\dfrac{x-1}{7} + \dfrac{x-2}{10} = x - 17$.

4. $\dfrac{3x+7}{4} + \dfrac{3x-15}{6} = \dfrac{15x-8}{5} - 6\dfrac{11}{15}$.

5. $\dfrac{3\,(2x-1)}{7} - \dfrac{4\,(x+3)}{5} + 3\dfrac{1}{7} = \dfrac{3}{5}$.

6. $\dfrac{3\,(x+7)}{5} - \dfrac{5}{9}\,(2x-1) = \dfrac{2}{3}$.

7. $x + \dfrac{4}{5} - \dfrac{x-3}{2} = 6\dfrac{3}{10}$.

8. $\dfrac{3x+7}{2} - \dfrac{4x-5}{3} = 6$.

9. $\dfrac{2x+1}{5} - x + 24\tfrac{1}{3} = \dfrac{x-2}{3}.$

10. $\dfrac{3(3x-4)}{7} - \dfrac{x-6}{3} + \dfrac{2(x+9)}{21} = 7\tfrac{3}{7}.$

11. $\dfrac{3(x+5)}{5} - \dfrac{4(5x-3)}{15} + \dfrac{2}{15} = 1.$

12. $\dfrac{x-3}{2} + \dfrac{1}{6} + \dfrac{x-4}{3} = 5\tfrac{1}{6} - \dfrac{8(x-4)}{12}.$

13. $\dfrac{4-5x}{6} - \dfrac{1-2x}{3} = \dfrac{13}{42}.$

14. $\dfrac{25-x}{2} - \dfrac{x}{7} = \dfrac{3x-25}{7}.$

15. $\dfrac{2(x-3)}{3} - \dfrac{6-2x}{12} = \dfrac{x+2}{6} - \dfrac{5(5-x)}{12}.$

64. Problems leading to simple equations.

Before solving a problem the facts stated in it must be set down in the form of an algebraic statement.

Some letter or symbol is chosen to represent that which is required to be found, and round this letter or symbol is built up the algebraic statement in the form of an equation, which being solved gives the result required.

EXAMPLE 1. If 4 be added to a certain number, the result is three times the number. Find it.

Let $x =$ the required number.

Then $(x+4) =$ the number with 4 added. The statement says that this is 3 times the number.

$$\therefore \quad 3x = x+4.$$
$$\therefore \quad 3x - x = 4,$$

whence $\qquad x = 2.$

EXAMPLE 2. Two people share £50 between them, so that the sixth part of the larger share is £4 more than the seventh part of the smaller share. Find the shares.

Let $\qquad x =$ the larger share.

Then $\qquad (50-x) =$ the smaller share.

Now $\dfrac{x}{6}$ = the sixth part of larger share,

and $\dfrac{(50-x)}{7}$ = the seventh part of smaller share.

$\therefore \quad \dfrac{x}{6} - \dfrac{50-x}{7} = 4.$

$7x - 6(50-x) = 168,$

$7x - 300 + 6x = 168,$

whence $\qquad 13x = 468.$

$\therefore \quad x = 36.$

$\therefore \quad 60 - x = 14.$ Thus the shares are £36 and £14.

Exercises 29

1. Find three consecutive *whole* numbers whose sum is 72.

2. Find three consecutive *odd* numbers whose sum is 51.

3. What number multiplied by 3 and then diminished by 14 will leave 7?

4. Find a number whose third part exceeds twice its seventh part by 5.

5. If three consecutive *even* numbers are divided by 4, 5 and 6 respectively the sum of the quotients is 6. Find the numbers.

6. Divide 10s. between two people, so that one-third of the smaller share will equal one-fifth of the larger share.

7. A man sold a lathe for £12 and half as much as he paid for it. He made a profit of £2. 10s. Find the cost price.

8. Think of a number, double it, add 4, divide the result by 3, and subtract one-fifth of the original number. The result is equal to one-half the original number thought of added to $\frac{1}{2}$. What is the number?

9. One tank contains 35 gall. and another 7 gall. Equal quantities are pumped into each, until one contains 3 times as much as the other. How much was pumped in?

10. A certain number of facing bricks were bought at $1\frac{1}{2}d.$ each; 15 were found to be broken, and the remainder were sold at 2d. each, making a total profit of 2s. 6d. How many were bought originally?

11. A man bought 50 articles, some at 2d. each and some at 3d. each. How many of each kind were bought if he spent altogether 10s. 5d.?

12. A workman's wages are raised 5*s.* a week. His wages now for 10 weeks amount to 5*s.* less than they previously did for 11 weeks. What are his new wages?

13. A father is now 40 years of age, and his son 15. In how many years will the father be twice as old as the son?

14. Four times A's age equals five times B's. Five years ago, three times A's age was equal to four times B's. What are their present ages?

15. The perimeter of a rectangle is 8 ft. Another rectangle twice as long, and half as broad is $3\frac{1}{2}$ ft. more in perimeter. Find the difference in their areas.

16. The length of a rectangle is 12 ft. Were it 3 ft. wider, its area would be 228 sq. ft. Find its width.

17. A person collected £21, partly in florins and partly in half crowns. There were 192 coins. How many of each?

18. Divide 182 oranges among A, B and C, giving B 10 more than C and A 12 more than B.

19. A man bought 2 articles, one costing 1*s.* 6*d.* more than the other. If he had paid 9*d.* less for each article, his total outlay would have been $\frac{5}{6}$ of what it was. Find the cost of each article.

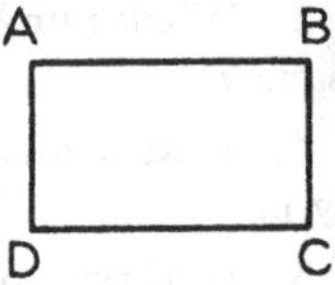

20. $AB = 8x - 2$; $DC = 5x + 10$; $BC = x^2 - 1$.
Find the area of the rectangle $ABCD$ in Fig. 11.

Fig. 11

65. Change of subject.

In the formula $LB = A$, used in para. 41, it will be equally true to express it as $\dfrac{A}{L} = B$ or $\dfrac{A}{B} = L$. In the original form A is expressed in terms of L and B, in the second case B is expressed in terms of A and L, while in the third case L is expressed in terms of A and B. This alteration in the form of an expression is called **changing the subject**, the subject being the single term expressed in terms of the other quantities. Here the subject was changed first from A to B, and then from A to L. If an expression involves a fraction, it is advisable to simplify it, so that the fractional form disappears before any attempt is made to change the subject.

EXAMPLE 1. In the following expression change the subject from V to f and find its value when $V = 3$ and $s = 2$.
$$V^2 = 2fs.$$

Divide across by $2s$,

$$\frac{V^2}{2s} = f = \frac{9}{4} = 2 \cdot 25.$$

EXAMPLE 2. If $d = \frac{3Wl^2}{2T}$, change the subject to l.

(1) Clear of fractions,
$$2Td = 3Wl^2.$$

(2) Divide across by $3W$,
$$\frac{2Td}{3W} = l^2.$$

(3) Take square root,
$$\sqrt{\frac{2Td}{3W}} = l.$$

EXAMPLE 3. If $d = \frac{WL^3}{4BD^3E}$, change the subject to B, D and L.

(1) Clear of fractions,
$$4BD^3Ed = WL^3.$$

(2) For B, divide by $4D^3Ed$,
$$B = \frac{WL^3}{4D^3Ed}.$$

(3) For D, divide by $4BEd$,
$$D^3 = \frac{WL^3}{4BEd}.$$

(4) Take cube root,
$$D = L\sqrt[3]{\frac{W}{4BEd}}.$$

(5) For L, divide by W,
$$L^3 = \frac{4BD^3Ed}{W}.$$

(6) Take cube root,
$$L = D\sqrt[3]{\frac{4BEd}{W}}.$$

Exercises 30

1. Express T in terms of P, R and I, and find its value when $P = 200$, $R = 3$, $I = 12$.
$$I = \frac{PRT}{100}.$$

2. In the following make V the subject, and find its value when $A = 21 \cdot 6$, $H = 10$:
$$\frac{3V}{A} = H.$$

3. Change the subject from A to r and find the value of r when $A = 200$ and $\pi = 3 \cdot 14$:

$$A = \pi r^2.$$

4. If $t = 2\pi \sqrt{\dfrac{l}{g}}$, express l in terms of the other quantities.

5. In the following express G in terms of I, W and L and find its value when $I = 0 \cdot 05$, $L = 3 \cdot 6$, $W = 0 \cdot 0115$:

$$\frac{WG}{L} = I + 0 \cdot 0497.$$

6. Change the subject from l to n, and find the value of n when $E = 15$, $r = 1\frac{2}{3}$, $l = 200$:

$$l = \frac{3Er}{10n}.$$

7. In the following make R the subject:

$$P = \frac{2\pi nR \, (P - W)}{H}.$$

8. Make T the subject of the following expression, and find its value when $a = 1\frac{1}{7}$, $R = 56$, $\pi = 3\frac{1}{7}$:

$$a = \frac{4\pi^2 R}{T^2}.$$

9. In the following express s in terms of F, D, p, d, and find its value when $F = 4480$, $D = 11\frac{4}{7}$, $p = 45$, $d = 24$:

$$F = \frac{psd^2}{D}.$$

10. Express the following in terms of M and find its value when $K = 27$, $S = 2$, $D = 21$, $\pi = 3\frac{1}{7}$:

$$K = \frac{\frac{1}{9}\pi^2 D^2}{2M^2 S}.$$

11. If $M = \frac{1}{2}V^2 \left(m + \dfrac{rs}{4}\right)$, find V in terms of the other quantities. What is its value when $s = 30$, $r = 1 \cdot 6$, $m = 20$, $M = 16$.

12. In the following change the subject to R and find its value when $V = 23\frac{4}{7}$, $h = 3$, $\pi = 3\frac{1}{7}$:

$$V = \frac{\pi h}{2} R^2 + \frac{\pi h^3}{6}.$$

CHAPTER VII

ANGLES AND ANGLE MEASURE

66. Consider Fig. 12(a). Imagine the line AB fixed whilst AC is free to rotate about A as a hinge. We say "AC has turned about A so as to sweep out the angle BAC".

Now consider Fig. 12(b). Imagine that in the first place AC started from the line AB and that it is able to turn about A. Imagine it to do this and we observe that, *as the amount of turning increases so also does the angular space between AB and AC.*

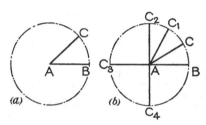

Fig. 12

The point C arrives successively at the positions C_1, C_2, C_3, C_4 and, having made one complete revolution, arrives back in its original position. The path of C is clearly a circle.

67. The common unit of angle measure.

The circle swept out by the complete rotation of AB about A (Fig. 12(b)) we may divide into the four equal parts or quadrants shown. We call each of the angular spaces BAC_2, C_2AC_3, C_3AC_4, C_4AB, a **right angle**. For convenience of comparison and measurement we divide a right angle into 90 equal **angular spaces known as degrees.**

Now observe that the right angle C_2AB is faced (or "subtended") by the arc C_2C_1CB. An angle of one degree, then, will be subtended by one ninetieth of this arc—or one three-hundred-and-sixtieth part of the whole circumference. Thus we arrive at a clear-cut definition.

A degree is the angle at the centre of a circle subtended by

an arc equal to one three–hundred–and–sixtieth part of the whole circumference.

Consider the angle *AOB* (Fig. 13). The straight lines *AO* and *BO* are known as *arms* whilst the point *O* is known as the *vertex*. We may more shortly describe this angle as $\angle AOB$ or $\hat{AOB}$.

The magnitude of $\angle AOB$ may be described as the amount of turning that must be given to *OA* to make it coincide with *OB*.

68. Kinds of angles (Fig. 13).

Acute angles: less than 90° ($\angle AOB$ in Fig. 13).

Obtuse angles: greater than 90° but less than 180° ($\angle COD$ in Fig. 13).

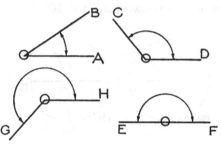

Fig. 13

Straight angles: equal 180° ($\angle EOF$ in Fig. 13).

Reflex angles: greater than 180° but less than 360° ($\angle GOH$ in Fig. 13).

Adjacent angles lie on either side of a common arm.

Thus in Fig. 14 the angles *BOC* and *AOB* lie on either side of the common arm *BO* and are adjacent.

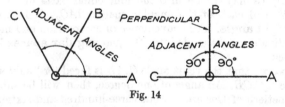

Fig. 14

Fig. 14 also shows a particular case where two adjacent angles are formed by one straight line *BO* standing on another straight line *AC*.

O is the common arm and the adjacent angles AOB and BOC are equal.
n such a case BO is said to be perpendicular to AC.

Complementary Angles: When two angles together make 90°
hey are said to be complementary. Thus 30° is the complement of 60°.

Supplementary Angles: When two angles together make 180°
hey are said to be supplementary. Thus 140° is the supplement of 40°.

69. The set–squares.

Most students will be aware that the set-squares in everyday use
are of two forms. In the first the angles are 90°, 45°, 45°; in the

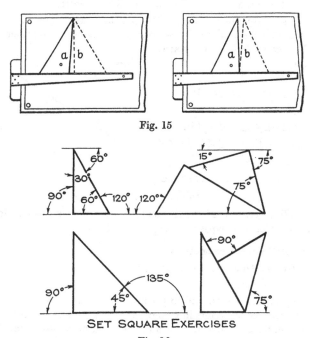

Fig. 15

SET SQUARE EXERCISES

Fig. 16

second 90°, 60°, 30°. Notice in passing that the three angles in each
add up to 180°.

Testing the right angle. Place the set-square on the edge of the
T-square—as in Fig. 15—and draw a perpendicular towards the top of

5-2

the paper. Now turn the set-square into position *b* and draw a line similar to the first. The two lines should coincide or be parallel. In the illustration the set-squares have exaggerated errors.

Set–square exercises (Fig. 16).

It may be useful here to mention that in general technical drawing the set-squares are used both singly and together to obtain angles other than 90°, 60°, 45°, 30°. A few cases are shown in Fig. 16. The student should endeavour to think of others, draw angles in this way, and test the results with the protractor.

70. Angles both supplementary and adjacent.

Let two straight lines *AB* and *CD* cross at *O* as in Fig. 17.

Then angles *AOD* and *DOB* are adjacent. Also they together form a straight angle *AOB*, i.e. they are supplementary.

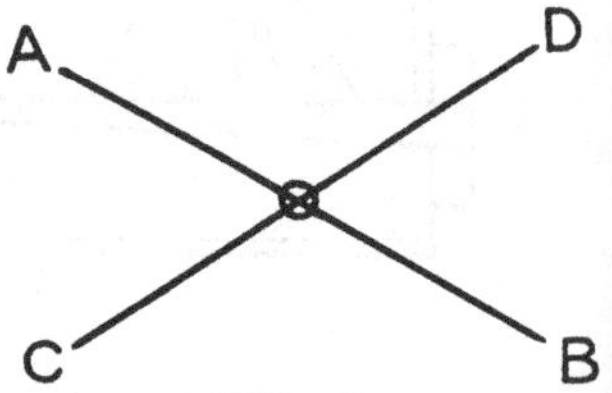

Thus we may write

$$\angle AOD + \angle DOB = 180°.$$

Similarly,

$$\angle AOD + \angle AOC = 180°.$$
$$\therefore \quad \angle AOD + \angle DOB = \angle AOD + \angle AOC.$$
$$\therefore \quad \angle DOB = \angle AOC.$$

Fig. 17

These angles are called **vertically opposite angles**.

Examples: $\angle AOC$ and $\angle DOB$; $\angle AOD$ and $\angle COB$.

Learn: **When two straight lines cross each other the vertically opposite angles are equal.** (*Proposition* 1.)

71. Simple angle exercises.

Ex. 1. *Given*: An angle (say $\angle AOB$, Fig. 18(*a*)).

Reqd.: To bisect it.

Method: With centre *O*, and any convenient radius, describe an arc cutting the arms of the angle in *C* and *D*. With *C* and *D* as centres, and any convenient radius, describe arcs intersecting in *E*. Join *OE*. Then *OE* bisects the angle $\angle AOB$.

Ex. 2. *Given*: A straight line (say *AO*, Fig. 18 (*b*)).

Reqd.: To draw a line *OE* at 60° to it.

Method: With centre *O* and any convenient radius (say *OB*) describe an arc. With centre *B*, and the *same* radius, describe an arc cutting the

ormer arc in *C*. Draw *OE* through *C*. Take your protractor and 60°
et-square and verify the accuracy of your construction.

Ex. **3.** *Reqd.*: To obtain angles of 30° and 15° (Fig. 18 (c)).

Method: Construct an angle of 60° as in the preceding exercise and by
isection obtain angles of 30° and 15°.

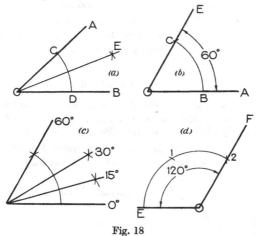

Fig. 18

Ex. **4.** *Reqd.*: To construct an angle of 120° (Fig. 18 (d)).

Method: The angle *EOF* in the illustration is clearly composed of
wo angles of 60°, each obtained as in Ex. 2.

Exercises 31

1. Using both set-squares obtain angles of 15°, 75°, 105°, 120°, 150°.
heck your accuracy.

2. Draw 5 lines radiating from a point as in
Fig. 19. Estimate the magnitude of each of the
ngles *A*, *B*, *C*, *D*, *E*. Then measure each with the
rotractor. Compare the results. Which are acute?
Which obtuse? What is their sum?

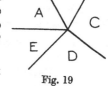

Fig. 19

3. What is the angle between the hands of a clock
t 3 o'clock, 6 o'clock, 9 o'clock, 12 o'clock?

4. Calculate the angle turned through by the minute hand of a clock
1 35 min.

5. Write down the supplements of 30°, 60°, 92°, 179°. Also give the complements of 21°, 45°, 75°, 81°.

6. Two straight lines *LM* and *NP* intersect at *O*. If the $\angle LON = 40°$, what is the magnitude of each of the angles *LOP*, *POM*, *MON*?

7. At what rate, in degrees per minute, does the minute hand of a clock move? At what rate does the hour hand move?

8. The circular card on a mariner's compass has its circumference divided into 32 equal parts. What is the magnitude of the angle subtended at the centre of the circle by each of these arcs?

PARALLELISM

72. Examine the cube in Fig. 3. Better still, examine an actual cubical object. Were you asked to make a cube—say of steel and having a half-inch edge—you would "file the opposite faces parallel".

Direction. (a) In Fig. 20 the line *OA* is drawn from *O* to *A* on a plane surface and throughout its length it has a fixed direction.

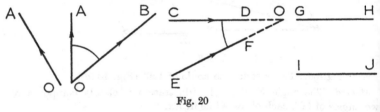

Fig. 20

(b) The lines *OA* and *OB* are each drawn from *O* but in different directions. The $\angle AOB$ measures the difference in their directions.

(c) The lines *CD* and *EF* converge towards a point *O*. The $\angle COE$ measures the difference in their directions.

(d) The lines *GH* and *IJ* have *like directions*, i.e. *they would never meet* however far they were produced. These lines are said to be parallel.

Ex. 1. *Given*: A straight line *EF*.

Reqd.: Parallels *GH* and *IJ*, 1½ in. from it.

Method: Draw *EF* as in Fig. 21. With any centre in *EF* (say point marked 3), describe arcs of 1½ in. radius on both sides of it. Do precisely the same from another point in *EF*. Then *GH* and *IJ*, which

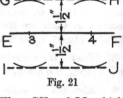

Fig. 21

ust touch the arcs, are the required parallels. Lines which "touch" or "meet" arcs are called *tangents*.

Ex. 2. Fig. 22 shows the plan of an iron plate. The sides AB and CD are parallel. Draw this view showing construction clearly.

73. Parallel lines and transversals.

Let AB and CD in Fig. 23 be two parallel lines cut by the transversal PS in Q and R.

1. *Corresponding angles.*

It is evident that parallel lines, having the same direction, must make equal angles with the transversal.

Thus $\angle a = \angle e$; $\angle d = \angle h$; $\angle b = \angle f$; $\angle c = \angle g$.

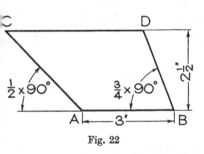

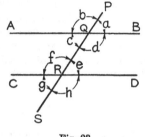

Fig. 22   Fig. 23

Considered in pairs these are called *corresponding angles*. They occur in similar relative positions and on the same side of the transversal.

2. *Alternate angles.*

$\angle a = \angle c$ (being vertically opposite), $\therefore$ $\angle c = \angle e$,

$\angle b = \angle d$ „   „   „ $\therefore$ $\angle d = \angle f$.

Considering these in pairs,

angles c and e; angles d and f, are called *alternate angles*.

3. *Interior angles on the same side of the transversal.*

Angles d and e, also angles c and f, are called interior angles; whilst angles a and b, also angles g and h, are called exterior angles.

Now $\angle d + \angle e = \angle c + \angle f = 2$ right angles.

(Since angles d and c are supplementary and angles c and e are alternate.)

Summarising. When two parallel straight lines are cut by a transversal,

(1) corresponding angles are equal,

(2) alternate angles are equal,

(3) interior angles on the same side of the transversal are supplementary.

74. *Problem*: Given a straight line *RS* (of any length) it is required to divide it into any number of equal parts (say 5) by means of set-square and ruler (Fig. 24).

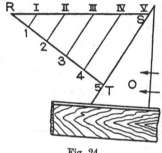

Fig. 24

Method. Draw *RT* of any length and making any angle with *RS*. Set off along it 5 equal parts of any convenient length. Join *TS*. Place *any* edge of *any* set-square along *TS* and, sliding it along a ruler, draw parallels 5v, 4iv, 3iii, 2ii, etc. Point out the corresponding angles on this diagram.

TRIANGLES (*Contd.*)

75. Kinds of triangles.

(1) Triangles may be classified by thinking of their angles, e.g.

  (*a*) *right-angled*, when one angle is a right angle.
    Example: △*ABC* in Fig. 25.

  (*b*) *obtuse-angled*, when one angle is obtuse.
    Example: △*DEF* in Fig. 25.

  (*c*) *acute-angled*, when *all* the angles are acute.
    Example: △*GHI* in Fig. 25.

(2) Triangles may be classified by thinking of their sides, e.g.

  (*a*) *equilateral*, have all sides equal.
    Example: △*JKL* in Fig. 25.

  (*b*) *isosceles*, have two sides equal.
    Example: △*MNO* in Fig. 25.

  (*c*) *scalene*, have all sides unequal.
    Example: △*PQR* in Fig. 25.

As a triangle has three sides it has three angular points. Any one of these angular points may be called a *vertex*, and the opposite side is then called the *base*. The *hypotenuse* of a right-angled triangle is the longest side and is situated opposite to the right angle. The line joining an angular point of a triangle to the mid-point of the opposite side is called a *median*. The three medians intersect at the *centroid*.

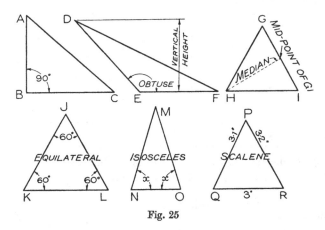

Fig. 25

76. Interior angles of a triangle.

Proposition 2. **The sum of the interior angles of a triangle is two right angles.**

Experimental verifications.

(1) Draw any scalene triangle, choosing bold sizes. For instance, it will be convenient to draw the sides 4 in., 4·2 in., 4·3 in.

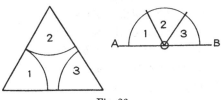

Fig. 26

Describe arcs of $1\frac{3}{4}$ in. radius from each of the angular points of the triangle in turn. You will now have the sectors 1, 2, 3 (see Fig. 26) which should be cut out carefully. Draw a straight line AOB (say $4\frac{1}{2}$ in. long) and fit the coloured or marked sectors together as in the illustration. State your conclusion.

(2) Draw any four triangles. By means of a protractor carefully measure the interior angles of each.

Complete the table:

No. of triangle	$\hat{A}$	$\hat{B}$	$\hat{C}$	$A+B+C$
1				
2				
3				
4				

Proof. If the side BC of the $\triangle ABC$ is produced to D, the $\angle ACD$ is called an exterior angle of the triangle (see Fig. 27).

The angles p and q are called interior and opposite angles.

Through C draw CE parallel to AB.

Since BCD meets the parallels BA and CE,

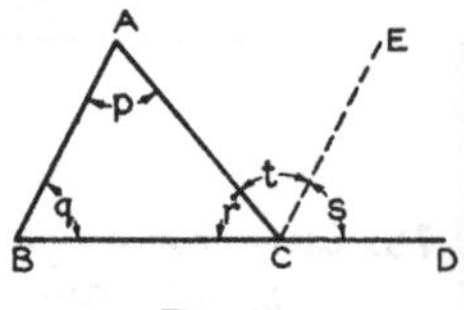

Fig. 27

$$\angle q = \angle s \quad \text{(corresponding angles)}.$$

Since AC meets the parallels AB and EC,

$$\angle p = \angle t \quad \text{(alternate angles)}.$$

Adding, $\qquad\qquad \angle q + \angle p = \angle s + \angle t,$

$$\therefore \ \angle ACD = \angle p + \angle q \qquad \dots \dots (1).$$

Adding $\angle ACB$ to each side of the equation,

$$\therefore \ \angle ACD + \angle ACB = \angle ACB + \angle q + \angle p.$$

But $\qquad\qquad \angle ACD + \angle ACB = 2 \text{ right angles}.$

$$\therefore \ \angle ACB + \angle p + \angle q = 2 \text{ right angles} \qquad \dots \dots (2).$$

Conclusion.

(1) *An exterior angle of a triangle is equal to the sum of the interior and opposite angles.*

(2) *The three interior angles of a triangle together make 2 right angles.*

(3) *The four interior angles of any quadrilateral together make* right angles—*for any quadrilateral can be divided into 2 triangles.*

Exercises 32

1. Draw two parallel straight lines AB and CD, and a third straight line EF crossing them as in Fig. 28. Describe each of the angles numbered 1 to 8. State which are equal and which are supplementary.

2. Suppose $\angle 2$ in Fig. 28 is $42°$. Write down the values of all other angles on the diagram.

3. Divide a line $3 \cdot 75$ in. long into 7 equal parts by means of ruler and set-square.

4. In a $\triangle ABC$, $\angle A = 60°$, $\angle B = 30°$. What is the magnitude of $\angle C$?

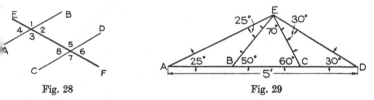

Fig. 28   Fig. 29

5. In a $\triangle ABC$, $\angle A = 70°$, and the exterior angle at $C = 130°$. What is $\angle B$?

6. In an isosceles $\triangle EFG$, $EF = FG$; $\angle F = 40°$. Calculate the magnitude of the exterior angles at E and G.

7. PQR is an isosceles triangle having $PQ = QR$. PQ is produced to N. Prove that $\angle NQR$ is twice $\angle PRQ$.

8. Construct a $\triangle EBC$ having a perimeter of 5 in. and base angles $40°$ and $60°$.
Fig. 29 should be consulted and the method analysed.

9. The angles X and Y of the $\triangle XYZ$ are respectively $60°$ and $50°$. The exterior angles at Z and Y (formed by producing the base YZ in both directions are bisected by ZO and YO. What is the magnitude of $\angle ZOY$?

77. Interior angles of polygons.

Fig. 30 shows a quadrilateral $ABCD$ and an irregular hexagon $EFGHIJ$.

Quadrilateral. Draw *DB* a diagonal. It joins opposite *angular points* or *vertices B* and *D*, thus dividing the quadrilateral into two triangles, viz. *DBA* and *DBC*.

Interior angles of quadrilateral

  = sum of interior angles of both triangles.
  = **4 right angles (360°).**

Hexagon. From *E* draw diagonals *EG, EH, EI,* thus dividing the figure into 4 triangles.

Interior angles of hexagon

  = sum of interior angles of 4 triangles.
  = **8 right angles (720°).**

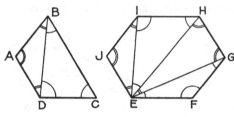

Fig. 30

By drawing other polygons and dividing them similarly it will be evident that the number of triangles so obtained is always 2 less than the number of sides of the polygon, i.e. if the polygon has n sides, there are $(n-2)$ triangles.

 ∴ **Sum of interior angles of any polygon**

  = $(n-2) \times 2$ right angles = **2n − 4 right angles.**

EXAMPLE. Interior angles of a nonagon '9 sides)

  = $9 \times 2 - 4 = 14$ right angles = $14 \times 90 = 1260°.$

78. Exterior angles of polygons.

Suppose *ABCDE* (a pentagon) in Fig 31 has its sides produced *in order*, thus forming the exterior angles a, b, c, d, e.

Choose any point *O* and from it draw radiating lines each parallel to a side of the polygon so as to obtain angles a, b, c, d, e, which evidently total 360°.

No matter what convex polygon be selected the result remains the same.

Conclusion. When the sides of a polygon are produced in order round the figure the sum of the exterior angles so formed is 360°.

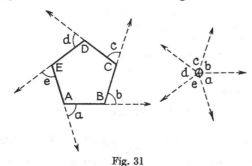

Fig. 31

79. Regular polygons.

1. *Interior angles.* Regular polygons have all their sides and all their angles equal. Consider a regular pentagon (Fig. 32).

Sum of interior angles $=(2n-4)$ right angles (para. 77)
$=6$ right angles $=540°$.

∴ Each interior angle $=\frac{540}{5}=108°$ (*see* $\angle b$ *in Fig. 32*).

∴ Each exterior angle $=180-108=72°$ (*see* $\angle a$ *in Fig. 32*).

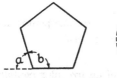

Fig. 32

Expressing this generally, if the polygon has n sides, the interior angle

$$=\frac{2n-4 \text{ (right angles)}}{n}=\left(2-\frac{4}{n}\right) \text{ right angles.}$$

2. *Exterior angles.* Having obtained the interior angle, as shown above, the exterior angle may be calculated thus:

$$180° - \text{interior angle} = \text{exterior angle.}$$

Another method is as follows:

Sum of exterior angles $= 360°$ (para. 78).

$$\therefore \quad \textbf{Any exterior angle} = \frac{\textbf{360}}{n}.$$

Note also that if any regular polygon be inscribed within a circle and lines drawn from the centre to any vertex we divide the polygon into as many isosceles triangles as the figure has sides.

In the case of the pentagon shown in Fig. 32,

$$\angle a + \angle b + \angle c + \angle d + \angle e = 360°.$$

But these 5 angles are equal,

$$\therefore \quad \text{Angle at centre} = \frac{360}{5} = 72°.$$

Expressing this generally, the magnitude of the angle at the centre of any regular polygon is $\frac{360}{n}$, where $n = $ no. of sides.

Exercises 33

1. State the sum of the interior angles of a regular pentagon, hexagon, heptagon, octagon.

2. State the magnitude of an exterior angle of every polygon mentioned in Question 1.

3. Three interior angles of a quadrilateral are 30°, 90°, 80°. Find the fourth.

4. If four of the interior angles of a regular pentagon are

$$a°, a+1°, a+2°, a+6°,$$

find the magnitude of the fifth.

5. Three of the interior angles of an irregular pentagon respectively equal 110°. If the fifth and sixth are equal, find the magnitude of each.

6. XY is the base of a regular nonagon. Every vertex is joined to the centre O. What is the magnitude of $\angle XOY$?

If XY is produced to Z, what is the magnitude of $\angle OYZ$?

80. Congruence of triangles.

We have seen that triangles have three sides and three angles, i.e. six *elements*. In geometrical drawing, in the construction of triangles from given particulars, 3 of the 6 elements are usually iven and in many cases the 3 given sizes enable one (and *only one*) riangle to be constructed. Sometimes, however, the information is o given that more than one correct solution could result. We then have what are called *ambiguous cases*. For example, suppose we re given the three angles of a triangle it is evident that any number f solutions could be drawn. All the resulting triangles would have the ame *shape*, but their *sizes* (i.e. sides and areas) could differ greatly.

When it is required to construct a triangle having definite *size* as well as *shape* we must provide information as follows:

Case 1. *Give two sides and the included angle.*
Case 2. *Give one side and two angles.*
Case 3. *Give the three sides.*
Case 4. *Give, in the case of a right-angled triangle, the hypotenuse nd one other side.*

In Fig. 33 the given elements are marked on each triangle.

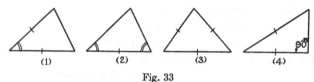

Fig. 33

Exercises 34

Construct triangles given the following information:

1. Sides 3 in., 2 in. Included $\angle 45°$.
2. Base $3\frac{1}{4}$ in. Base angles 35°, 45°.
3. Sides 3 in., 3·3 in., 3·9 in.
4. Right-angled triangle. Base 3 in., hypotenuse $4\frac{1}{2}$ in.
5. Base angle 40°. Sides 3·45 in., 2·4 in. (This will yield an ambiguous esult.)

81. What are congruent triangles?

When two triangles have 6 elements of the one equal, respectively, o 6 elements of the other, they are obviously **equal in all respects**.

Thus, examine the triangles ABC nd RST (Fig. 34).

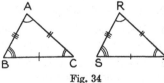

$AB = RS$; $AC = RT$; $BC = ST$.
$\angle A = \angle R$; $\angle B = \angle S$; $\angle C = \angle T$.

Fig. 34

Suppose these triangles be cut out and the $\triangle ABC$ superimposed on $\triangle RST$ (or *vice versa*), one will coincide exactly with the other and we should call them congruent triangles.

If we draw two triangles from the particulars given in *any one* of the cases 1, 2, 3, 4 (para. 80), we shall have a pair of congruent triangles. In other words, if two triangles have three elements equal, each to each, as set out in any one of the cases 1, 2, 3, 4, it follows that the remaining elements in any such pair are equal also.

82. *Proposition* 3. **In an isosceles triangle the angles opposite the equal sides are equal.**

Let ABC (Fig. 35) be an isosceles triangle in which $AB = AC$.
It is required to prove that $\angle B = \angle C$.

Construction. Draw AD bisecting $\angle A$ and meeting BC in D.

Proof. In the triangles ABD and ACD,

$$AB = AC \quad \text{(given)},$$
$$AD = AD \quad \text{(common to both)}.$$

Included $\angle BAD$ = included $\angle CAD$
(by construction).

∴ The triangles are congruent (Case 1, para. 80).
∴ The remaining three elements are equal.

$$\therefore \quad \angle B = \angle C.$$

Fig. 35

83. The converse of a proposition.

In the foregoing we were *given* $AB = AC$. We *proved* $\angle B = \angle C$.
The converse would be: given $\angle B = \angle C$, to prove side AB = side AC.

Exercises 35

1. Prove that if two angles of a triangle are equal, then the opposite sides are equal, i.e. the triangle is isosceles.

2. If a triangle is equilateral, prove that it is also equiangular.

3. Draw a straight line XY, bisect it at right angles. Prove that any point on this bisector is equidistant from X and Y.

4. In a triangle ABC the exterior angles at A and B are equal. Prove that the triangle is isosceles.

Exercises 36

1. As in Ex. 1, para. 71, draw an acute angle AOB and bisect it by line OE. Select any point R in OE and from it drop perpendiculars RQ, RS to OA and OB respectively. Prove that $RQ = RS$.

2. PQR is an equilateral triangle. Along QR mark points S and T so that distances $QS = RT$. Prove that the triangle PST is isosceles.

3. Prove that if a perpendicular AD be dropped from the vertex A of an isosceles triangle ABC, it bisects the base BC.

4. $ABCDEF$ is a regular hexagon. Prove that $AC = BD$ and that the triangle BDF is equilateral.

CHAPTER VIII

QUADRILATERALS

84. A **quadrilateral** is a plane figure bounded by four straight lines.

A **parallelogram** is a quadrilateral whose opposite sides are parallel.

A **rectangle** is a parallelogram whose angles are right angles.

A **rhombus** is a parallelogram having all sides equal but its angles are not right angles.

A **square** is a rectangle having its adjacent sides equal.

A **trapezium** is a quadrilateral having two opposite sides parallel.

85. *Proposition* 4. **In any parallelogram (1) either diagonal divides it into two congruent triangles, (2) the opposite sides are equal, (3) the opposite angles are equal.**

Let $ABCD$ be a parallelogram in which AB is parallel to DC, AD is parallel to BC.

Draw a diagonal DB. See Fig. 36.

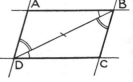

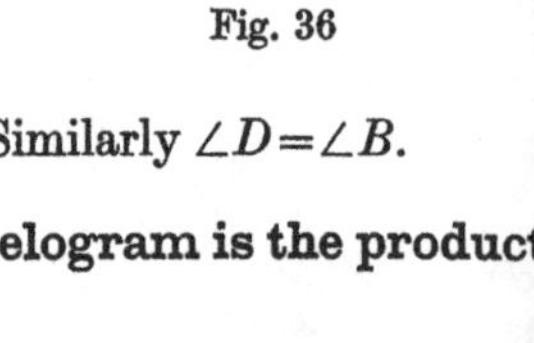

Proof: In the triangles ABD, CBD

$\begin{cases} DB \text{ is common to both.} \\ \angle ABD = \angle BDC \text{ (alternate angles).} \\ \angle ADB = \angle DBC \qquad ,, \qquad ,, \end{cases}$

∴ Triangles are congruent (Case 2, para. 80).

Fig. 36

∴ $AD = BC$, $AB = DC$ and $\angle A = \angle C$. Similarly $\angle D = \angle B$.

86. *Proposition* 5. **The area of a parallelogram is the product of its base and altitude.**

Let $ABCD$ be a parallelogram (Fig. 37).

Construction. Draw DE, CF perpendicular to DC, meeting BA (or BA produced) in E and F. Draw ST perpendicular to DC.

Proof: $DCFE$ is a rectangle (by construction).

∴ $DE = CF = ST$, and $EF = DC$ (prop. 4).

In the triangles EDA and FCB:

$$\left.\begin{array}{l} DE = CF \\ \angle DEA = \angle CFB \\ DA = CB \end{array}\right\}$$ ∴ Triangles are congruent. (Case 4, para. 80.)

∴ Triangles EDA and FCB are equal in area.

Add to each triangle the trapezium $ADCF$.

∴ △FCB + trapezium $ADCF$
$$= \triangle EDA + \text{trapezium } ADCF \ (\textit{in area}).$$

∴ Parallelogram $ABCD$
$$= \text{parallelogram } EDCF \ (\textit{in area}).$$

But parallelogram $EDCF$ is a rectangle.

∴ Area of parallelogram $ABCD = DC \times CF$,
$$= DC \times ST,$$
$$= \text{Base} \times \text{Altitude}.$$

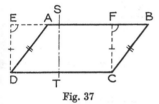

Fig. 37

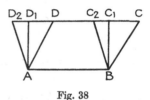

Fig. 38

87. *Proposition* 6. **Parallelograms standing on the same base and lying between the same parallels are equal in area.**

The proof of this follows from that in para. 86. In Fig. 38 the parallelograms $AD_2C_2B = AD_1C_1B = ADCB$ for they stand on the same base AB and lie between the parallels AB and CD_2.

88. *Proposition* 7. **The area of a triangle is half the product of base and altitude.**

Let PQR be a triangle and PS a perpendicular from P to QR. Then PS is the altitude. See Fig. 39.

Construction. Draw TP parallel to QR, TQ parallel to PR.

Proof. $TPRQ$ is a parallelogram (by construction).

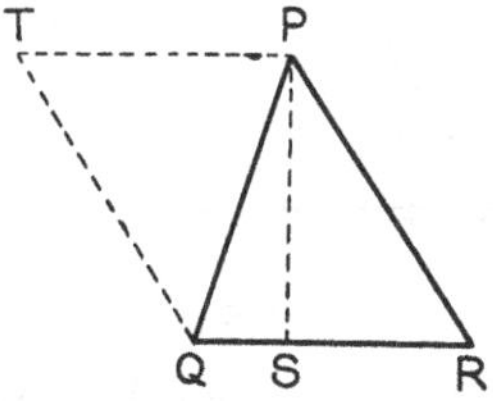

∴ QP bisects the parallelogram (prop. 4, para. 85).

Fig. 39

∴ Triangles TPQ and PQR are congruent and equal in area.

$$\therefore \quad \triangle PQR = \tfrac{1}{2} \text{ area of parallelogram } TPRQ,$$
$$= \tfrac{1}{2} QR \times PS,$$
$$= \tfrac{1}{2} \text{ Base } \times \text{Altitude}.$$

89. *Proposition* 8. **Triangles standing on the same base and lying between the same parallels are equal in area.**

The proof of this is left to the student. See Fig. 40.

Triangles ABC, ABC_1, ABC_2, ABC_3

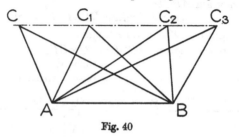

Fig. 40

stand on the same base AB and lie between the same parallels AB and CC_3.

90. Reduction of polygons to equivalent triangles.

Fig. 41. *Given.* Quadrilateral $ABCD$. *Reqd.* Triangle of equal area.

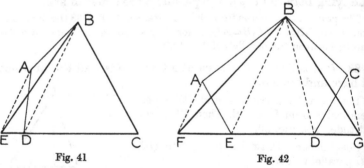

Fig. 41   Fig. 42

Method. Join two vertices (say B, D). Through A draw AE parallel to BD and meeting the base produced in E. Join BE. Then BEC is the required triangle.

Proof. Consider the triangles *BDA* and *BDE*. They stand on the same base *BD* and lie between the same parallels *BD* and *AE*.

∴ △*BDA* = △*BDE* in area. (Note that they are *not* proved to be congruent.)

To each add △*BDC*.

$$\therefore \quad \triangle BDA + \triangle BDC = \triangle BDE + \triangle BDC,$$

i.e. Quadrilateral *ABCD* = △*BEC*.

Fig. 42. *Given.* Pentagon *ABCDE*. *Reqd.* Triangle of equal area.

Method. This will be obvious. The proof is left to the student.

91. Area of a trapezium.

Let *ABCD* in Fig. 43 be a trapezium having sides *AB* and *CD* parallel. Join *AC* and drop perpendiculars from *A* and *B* to the base *DC*.

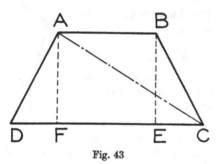

Fig. 43

Then area of *ABCD* = area of △*ADC* + area of △*ABC*

$$= \tfrac{1}{2} DC \times BE \quad + \tfrac{1}{2} AB \times BE$$
$$= \tfrac{1}{2} BE (DC + AB)$$
$$= \tfrac{1}{2} \textbf{ height} \times \textbf{ sum of parallel sides.}$$

EXAMPLE 1. Find the area of a trapezium having its parallel sides 45 ft. and 55 ft., the distance between them being 30 ft.

$$\text{Area of trapezium} = \tfrac{1}{2} \times 30 \times (45 + 55)$$
$$= 15 \times 100$$
$$= 1500 \text{ sq. ft.}$$

EXAMPLE 2. Find the height of a trapezium if its parallel sides are
20 in. and 25 in., and its area is 225 sq. in.

$$\text{Area} = \tfrac{1}{2}h \times (20 + 25)$$
$$= 22\tfrac{1}{2}h.$$
$$\therefore \quad 22\tfrac{1}{2}h = 225.$$
$$\therefore \quad h = 10 \text{ in.}$$

92. Area of any quadrilateral.

Let $ABCD$ in Fig. 44 be any quadrilateral. Draw a diagonal BD
and from A and C drop perpendiculars AF and CE. Such perpendiculars are called *offsets*.

Area of quadrilateral

$ABCD =$ area of $\triangle ABD +$ area of $\triangle BCD$
$\quad = \tfrac{1}{2}(DB \times AF) + \tfrac{1}{2}(DB \times CE)$
$\quad = \tfrac{1}{2}DB(AF + CE)$
$\quad = \tfrac{1}{2}$ **diagonal × sum of the offsets.**

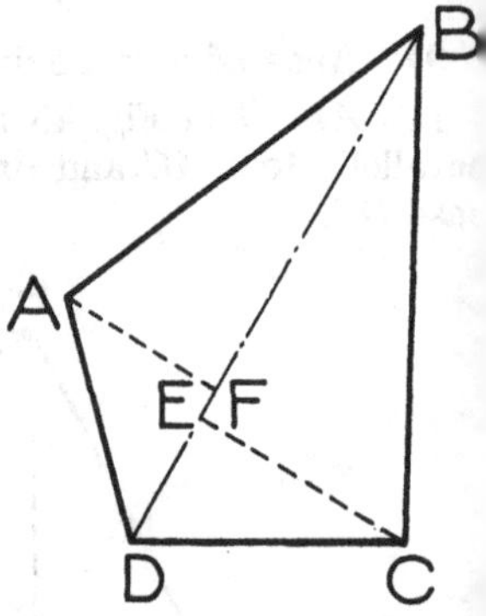

Fig. 44

Exercises 37

1. Construct a parallelogram whose area
is 10 sq. in., base 3 in., and one base angle
60°. State its vertical height.

2. On the same base construct two triangles each equal in area to the parallelo-
gram in Question 1, one having a base angle of 45°, the other having
a base angle of 50°.

3. Construct the following quadrilateral $ABCD$ by the method of
offsets. Find its area in square yards.

$CE = 140$ yd.; $CF = 160$ yd.; $CA = 180$ yd.; $BF = 120$ yd.; $DE = 100$ yd.
(*Scale* 2 *in.* = 100 *yd.*)

4. Construct a square on a base of 2 in.:

(a) Draw a triangle equal in area to the square but standing on a
base of $2\tfrac{1}{4}$ in.

(b) On the same base draw an isosceles triangle equal in area to
the triangle in (a).

5. Draw a regular hexagon on a base 2 in. Reduce it to a triangle of qual area. Measure the height and base of the triangle, and calculate ts area.

6. The angles A, B, C of a quadrilateral $ABCD$ are 84°, 96° and 140° espectively, and the bisectors of the four exterior angles of this quadilateral are drawn, forming the quadrilateral $PQRS$. Find the angles f this quadrilateral. Fig. 45 will suggest a method of solution. Note hat the angles A and B are supplementary, thus sides AD and BC are arallel.

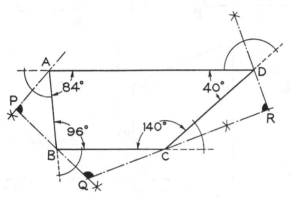

Fig. 45

7. (a) See Fig. 46 (a). Construct a triangle ABC. From D, any point n AC, draw DE parallel to AB, and meeting CB in E. Cross-hatch the rapezium $ABED$.

(b) Suppose that $AB = 4$ in., $DE = 2.75$ in., the perpendicular distance between them is 1·75 in. Find the area of the trapezium $ABED$.

8. A vertical dam of the section and dimensions shown in Fig. 46 (b) s formed across a river. Find the cross-sectional area of the dam.

9. Find the area of the cross-section of the irregular pentagon $ABCDE$ (Fig. 46 (c)). It represents a section of a motor-cycle garage, ides EA and CB being the same height.

10. Find the cross-sectional area of the mansard roof shown in Fig. 46 (d).

11   The area of a trapezium is 23·3 sq. in., and its height is $3\frac{1}{2}$in. On of the parallel sides is twice as long as the other. Find the length of bot. parallel sides.

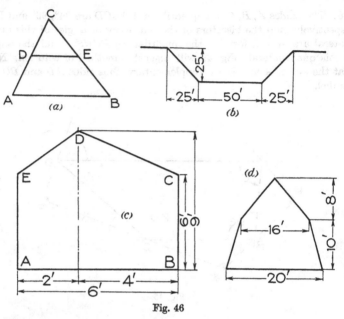

Fig. 46

93. Areas of regular polygons.

By joining the angular points of a polygon to the centre we ca divide it into as many isosceles triangles as there are sides to th polygon. Fig. 47 clearly shows that the radius of the inscribed circl is equal to the vertical height of the triangle.

The polygon chosen in Fig. 47 is a regular hexagon and the si: triangles of which it is composed have been re-arranged to form parallelogram—which thus comprises the *development* of the hexagon

Examine this parallelogram and note that:

(1) its base equals half the perimeter of the polygon;

(2) its height equals the radius (r) of the inscribed circle;

(3) its area = $\frac{1}{2}$ perimeter of polygon × r,

     or   = $\frac{1}{2}r$ × **perimeter of polygon.**

Suppose that a polygon having very many sides be circumscribed about a circle, then each side will form a very small tangent to the circle. If the sides are sufficiently numerous the perimeter of the polygon will hardly be distinguishable from the circumference of the circle. Thus we may consider a circle as a polygon having an infinite number of sides. This is referred to in para. 112.

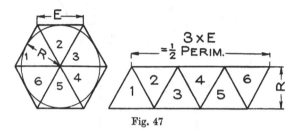

Fig. 47

Exercises 38

1. A regular hexagon is circumscribed about a circle of 2 in. radius. Its sides are each 2·309 in. long. What is its area?

2. Another rule for finding the area of a regular hexagon is

$$(\text{side})^2 \times 2{\cdot}598.$$

Find the area by this method and check your result.

94. The theorem of Pythagoras.

Fig. 48. Construct any right-angled triangle ABC, right-angled at C. The side AC opposite to B is denoted by b; side CB opposite to A is called a; side AB opposite to C is called c. The side c opposite to the right angle is called the *hypotenuse*.

Experiment 1. On the sides of the triangle construct squares. Find the mid-point O of the square on CB. Through this point draw one line parallel to, and one perpendicular to, the hypotenuse AB. Thus divide the square into four *congruent quadrilaterals*. Number these as shown. Now cut them out and superimpose them, with the square on AC, upon the square on the hypotenuse. It will be found that they exactly fit the square on AB.

The result suggests that

$$square \ on \ hypotenuse = square \ on \ AC + square \ on \ CB,$$
i.e.
$$c^2 = a^2 + b^2.$$

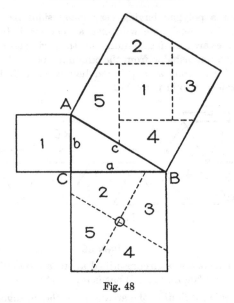

Fig. 48

Experiment 2. Draw a number of right-angled triangles having ∠C a right angle and carefully measure the sides a, b, c. Complete the following table.

△	a^2	b^2	a^2+b^2	c^2
1				
2				
3				
4				

Careful measurement and calculation will no doubt show that

$$c^2 = a^2 + b^2.$$

These, and similar *verifications*, point to a definite conclusion which we shall enunciate as the Theorem of Pythagoras. It is not only capable of graphical verification but also of geometrical proof.

The square on the hypotenuse of a right-angled triangle is equal to the sum of the squares on the other two sides.

EXAMPLE 1. The sides of a rectangular room are 30 ft. and 24 ft. What is the length of a diagonal drawn across the floor?

It is best to commence by making a rough sketch of the floor and inserting known dimensions upon it. The diagonal, of course, divides the rectangle into two right-angled triangles.

Let length of diagonal $= D$ ft.

Then
$$D^2 = 30^2 + 24^2$$
$$= 900 + 576$$
$$= 1476.$$
$$\therefore \quad D = \sqrt{1476} = 38 \cdot 42 \text{ ft.}$$

EXAMPLE 2. A flagstaff is 40 ft. high. A wire stay runs from the top of the pole to a point 20 ft. from the foot of the staff. Calculate the length of the stay (see Fig. 49).

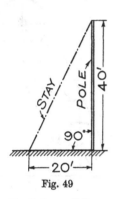

Fig. 49

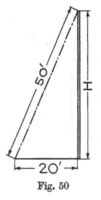

Fig. 50

Let length of stay $= L$ ft.

Then
$$L^2 = 20^2 + 40^2$$
$$= 400 + 1600$$
$$= 2000.$$
$$\therefore \quad L = \sqrt{2000} = 44 \cdot 72 \text{ ft.}$$

EXAMPLE 3. Suppose that the stay in the previous example is 50 ft. long. How high will be the flagstaff if the ground measurement, 20 ft., remains the same (Fig. 50)?

Let $H =$ height of flagstaff
$$50^2 = H^2 + 20^2.$$
$$\therefore \quad H^2 = 50^2 - 20^2$$
$$= 2100.$$
$$\therefore \quad H = \sqrt{2100} = 45 \cdot 82 \text{ ft.}$$

Important summary. (See Fig. 48.)

(1) $c^2 = a^2 + b^2$. $\therefore$ $c = \sqrt{a^2 + b^2}$.

(2) $a^2 = c^2 - b^2$. $\therefore$ $a = \sqrt{c^2 - b^2}$.

(3) $b^2 = c^2 - a^2$. $\therefore$ $b = \sqrt{c^2 - a^2}$.

Exercises 39

(The student will find it interesting to check his calculated answer by drawing and measurement.)

1. If the base of a right-angled triangle is 3 in., and its height is 4 in. find the length of its hypotenuse.

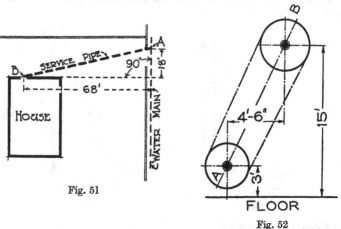

Fig. 51

Fig. 52

2. Using the same lettering as in Fig. 48, find the missing dimension in the following, which relate to right-angled triangles:

(1) $a = 2$ in.   (2) $a = 21$ cm.   (3) $a = 153$ ft.
    $c = 3$ in. $b = 28$ cm. $c = 185$ ft.
    $b = ?$. $c = ?$. $b = ?$.

3. From the information given in Fig. 51, find the length of the service pipe from the valve box at A to the stop cock at B.   (N.C.

4. Two squares of sheet lead with sides 7·5 cm. and 10 cm. respectively are melted down to make a single square of the same thickness. Find the length of the side of the new square.

5. Fig. 52 shows the details of a belt drive. Calculate the centre distance between the shafts.

6. The length and breadth of a rectangular building site are respectively $x + 16$ and x.

(a) Express the area of the site in algebraic terms.

(b) Express in algebraic terms the length of the diagonal of the site.

7. The length and breadth of a rectangular board are respectively $x - 2$ ft. and $x + 2$ ft. What would be the length of the side of a square board having the same area?

What would be the length of the diagonal of the square? (*N.C.*)

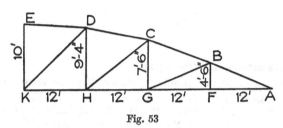

Fig. 53

8. *ABCD* is a rectangular field of which the sides *AB* and *CD* are 300 ft. long, while *BC* and *AD* are 240 ft. long. *E* is the mid-point of the side *CD*. Find, to the nearest foot, the length of the path from *A* to *E*.

9. The diagonals of a rectangle are 4·58 in. long. The length of one side is 4·42 in. long. Find the area.

10. Fig. 53 is a line diagram of one-half of a framed girder. Draw the figure to the dimensions given, using a scale of 1 in. = 10 ft. Calculate the lengths of the diagonals *BG*, *CH* and *DK*. Check results by measurement.

11. Draw the piece of sheet lead shown in Fig. 54, using a scale of $\frac{1}{4}$ in. = 1 ft. Calculate (a) its perimeter, (b) its area.

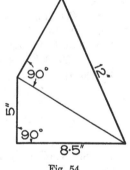

Fig. 54

12. Lines *XY* and *YZ*, 3 ft. 6 in. and 2 ft. 8 in. long respectively, are at right angles and lie in the horizontal plane. A point O is 2 ft. 2 in above Z. Calculate the distances OX and OY (see Fig. 55).

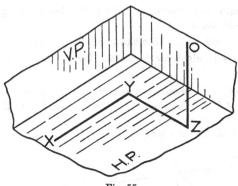

Fig. 55

95. The set–square triangles.

In the course of our practical calculations we shall frequently meet two right-angled triangles whose angles are 90°, 45° and 45° and 90°, 60° and 30° respectively. It is useful, therefore, to examine the ratios of the sides of these important triangles. We can then apply our knowledge to the solution of typical practical calculations.

(1) The triangle with angles 90°, 45°, 45°.

Let ABC (Fig. 56) be a right-angled triangle in which the sides CB and CA are each one unit in length.

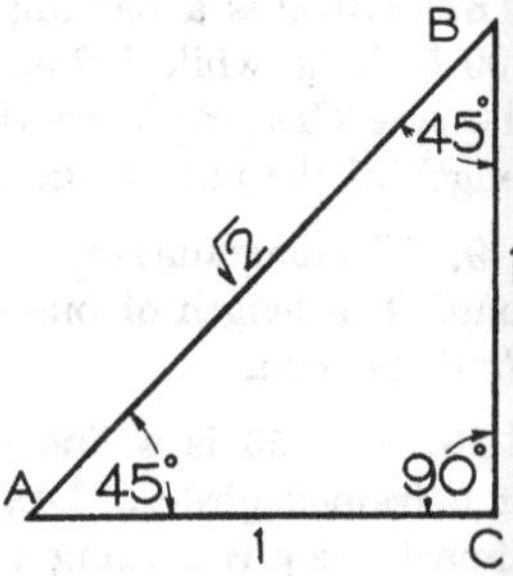

$\angle C = 90°$, $\angle A = \angle B = 45°$.

By the theorem of Pythagoras:
$$AB^2 = AC^2 + CB^2$$
$$= 1^2 + 1^2 = 2.$$
$$\therefore \quad AB = \sqrt{2}$$

(Note that $\sqrt{2} = 1\cdot414$ approx.).

$\therefore$ Sides AC, CB, AB are in the ratio **1 : 1 : $\sqrt{2}$**.

Thus
$$AB = \sqrt{2} \times AC; \ AB = \sqrt{2} \times BC; \ AC = \frac{AB}{\sqrt{2}}; \ BC = \frac{AB}{\sqrt{2}}.$$

Fig. 56

(2) The triangle with angles 90°, 60°, 30°.

Let ABC (Fig. 57) be an equilateral triangle. Join A to D, the mid-point of BC. Then triangles ADB and ADC are congruent, and $\angle ADB$ is a right angle.

By construction we know that

$$\angle ABD = 60°, \quad \angle BAD = 30°,$$
$$\text{side } BD = \tfrac{1}{2}AB.$$

Suppose BD is 1 unit in length. Then $AB = 2$ units.

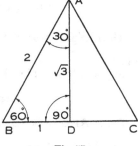

Fig. 57

Then

$$AD^2 = AB^2 - BD^2$$
$$= 2^2 - 1^2 = 3.$$
$$\therefore \quad AD = \sqrt{3}$$

(Note that $\sqrt{3} = 1.732$ approx.).

$\therefore$ Sides BD, BA, AD are in the ratio $1 : 2 : \sqrt{3}$.

Thus

$$AD = \sqrt{3} \times BD; \ AB = 2 \times BD; \ AD = \frac{\sqrt{3}}{2} \times BC.$$

Exercises 40

1. The sides of a square are 4 in. What is the length of its diagonal?

2. A square is inscribed within a circle. Its sides are 2·8 in. long. What is the diameter of the circle?

3. If the diagonal of a square is 6 in. long, what is the length of its side, and what is its area?

4. If the equal sides of a 45° set-square are 4 in. long. What is the length of the longest edge?

5. An equilateral triangle stands on a base of 4 in. Find its vertical height.

6. Fig. 58 shows a section of a vertical dam across a river. What is the length of its sloping sides?

Fig. 58

Fig. 59

7. Fig. 59 shows a section of an open water channel. What is its width at the top?

96. Area of an equilateral triangle.

Fig. 57 clearly demonstrates that an equilateral triangle is composed of two "30°–60°" triangles. We saw that

$$AD = \sqrt{3} \times BD, \quad \text{or} \quad AD = \frac{\sqrt{3}}{2} \times BC.$$

Thus height of equilateral $\triangle = \dfrac{\sqrt{3}}{2} \times$ base. (Note that $\dfrac{\sqrt{3}}{2} = 0.866$ approx.)

Suppose base $= a$ in.; height $= 0.866a$ in.

$$\text{Area} = \frac{a \times 0.866a}{2} = 0.433a^2.$$

Rule: Area of equilateral triangle = 0·433 × (side)².

97. Area of regular hexagon.

A regular hexagon can be divided into six equilateral triangles.
Thus
$$\text{area of hexagon} = 6 \times 0.433 \ (\text{side})^2 = 2.598 \ (\text{side})^2.$$

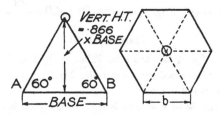

Fig. 60

Rule: Area of regular hexagon = 2·598 × (side)².

98. Finding square roots graphically.

To obtain graphically $\sqrt{2}$, $\sqrt{3}$, $\sqrt{4}$, $\sqrt{5}$.

Method. See Fig. 61(*a*). Draw a right-angled isosceles triangle having both sides about the right angle one unit in length. It will be clear that the area of the square on the hypotenuse is 2 sq. units and the length of the hypotenuse is $\sqrt{2}$. The rest of the construction is obvious.

Hints. (1) The theorem of Pythagoras applies only to right-angled triangles.

(2) $\sqrt{2} \times \sqrt{2} = 2$; $\sqrt{3} \times \sqrt{3} = 3$; $\sqrt{4} \times \sqrt{4} = 4$, etc.

In Fig. 61 (b) we show a simpler and more direct construction which should give a more accurate result inasmuch as it reduces the number of geometrical constructions.

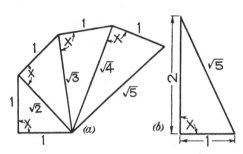

Fig. 61

99. Further examples—Pythagoras.

EXAMPLE 1. The area of a triangle ABC (shown in Fig. 62) is 1200 sq. yd. A pole is to be inserted along AC at a point P, so that BP is at right angles to AC. Calculate distance PC.   (N.C.)

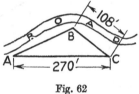

Fig. 62

Area of $\triangle ABC = \dfrac{b \times h}{2}$; $b = 270$ ft.

$$\therefore \quad \frac{b \times h}{2} = 1200 \times 9 \text{ (sq. ft.).}$$

$$\therefore \quad h = \frac{1200 \times 9 \times 2}{270} = 80 \text{ ft.}$$

Now consider the right-angled triangle BPC, in which $BP = 80$ ft., $BC = 108$ ft. (Sketch the triangle inserting the point P.)

Then
$$PC^2 = BC^2 - BP^2$$
$$= 108^2 - 80^2$$
$$= 5264.$$
$$\therefore \quad PC = \sqrt{5264} = 72 \cdot 6 \text{ ft.}$$

EXAMPLE 2. The length, breadth and height of a box are as shown in Fig. 63, dimensions being in feet. Find the length of the piece of string which will stretch from a corner (E) to the one diagonally opposite (F).

$$EG = \sqrt{5^2 + 3^2} = \sqrt{34}.$$

Now take right-angled triangle FGE in which $\angle FGE = 90°$.

$$EF^2 = EG^2 + FG^2$$
$$= (\sqrt{34})^2 + 6^2$$
$$= 34 + 36 = 70.$$
$$\therefore \quad EF = \sqrt{70} = 8 \cdot 37 \text{ ft.}$$

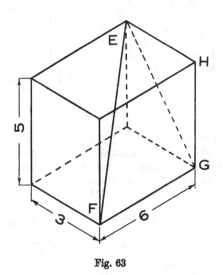

Fig. 63

Exercises 41

1. Each side and one diagonal of a rhombus is 10 in. long. Find its area.

2. An equilateral triangle has sides 8 in. long. Find (a) area, (b) length from vertex to base, (c) radius of circumscribing circle.

3. The floor of a summer house is a regular hexagon, each side measuring 10 ft. Find the cost of tiling it at 1s. 6d. per sq. ft.

4. Find the area of a regular hexagon the sides of which measure 3 in. If the hexagon is cut from a circular piece of wood 6 in. diameter, find the percentage area of the wood that is cut away.   (*N.C.*

5. The diagonal of a square baulk of timber is 14 in. long. What is the length of a side of the square?

6. Find the length of the diagonal of a rectangular prism the edges being 6 in., 8 in. and 10 in.

7. The sides of a triangle *ABC* are respectively 2 in., 10 in., 8 in. Find its altitude and thence its area. Take the base (*AB*) of the triangle as 2 in. long.

8. Calculate the cutting depth *x* necessary in machining a flat 1 in. wide on a shaft 3 in. diameter (see Fig. 64).

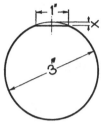

Fig. 64

CHAPTER IX

LOGARITHMS

100. In para. 35 the terms *power* and *root* were explained.
Since

$$a^1 \times a^1 = a^{1+1} = a^2, \qquad \therefore \quad \sqrt{a^2} = a,$$
$$\therefore \quad a^{\frac{1}{2}} \times a^{\frac{1}{2}} = a^{\frac{1}{2}+\frac{1}{2}} = a, \qquad \therefore \quad \sqrt{a} = a^{\frac{1}{2}},$$

that is, when the denominator of a fractional index is 2, the square
root of the quantity bearing that fractional index is indicated.

101. Quantities expressed to the power of 0.

By the index law
$$10^2 \div 10^2 = 10^{2-2} = 10^0.$$

But by ordinary division
$$10^2 \div 10^2 = 1.$$
$$\therefore \quad 10^0 = 1.$$
In general
$$a^0 = 1.$$

102. A new set of terms may now be introduced. In the expression
$10^2 = 100$, instead of saying that 100 is 10 to the power of 2, we say
that 2 is the **logarithm** of 100 to the **base** 10. In the expression
$10^3 = 1000$, 3 is called the logarithm of 1000 to the base 10. Any
number may be used as a base. Thus in $2^3 = 8$, 3 is the logarithm of 8
to the base 2. Logarithms which have 10 as the base are called
common logarithms. For the present common logarithms only
will be considered.

**The logarithm of a number is the power to which 10 is
raised to produce that number.**

103. Consider the following table:

$$
\begin{aligned}
10^3 &= 1000, & \therefore \quad \log \ 1000 &= \ \ 3. \\
10^2 &= \ 100, & \therefore \quad \log \ \ 100 &= \ \ 2. \\
10^1 &= \ \ 10, & \therefore \quad \log \ \ \ 10 &= \ \ 1. \\
10^0 &= \ \ \ 1, & \therefore \quad \log \ \ \ \ 1 &= \ \ 0. \\
10^{-1} &= \ \ 0 \cdot 1, & \therefore \quad \log \ \ \ 0 \cdot 1 &= -1. \\
10^{-2} &= \ 0 \cdot 01, & \therefore \quad \log \ \ 0 \cdot 01 &= -2. \\
10^{-3} &= 0 \cdot 001, & \therefore \quad \log \ \ 0 \cdot 001 &= -3.
\end{aligned}
$$

The logarithms of those numbers which lie between exact powers of 10, lie between the logarithms of these exact powers.

Thus
$$\log 900 = \quad 2 + \text{a fraction};$$
$$\log 0{\cdot}09 = -2 + \text{a fraction}.$$

The fractional part of a logarithm is always expressed as a decimal. The whole number part of a logarithm is called the **Characteristic** and the decimal part the **Mantissa**. The characteristic of a logarithm can always be found by inspection.

Examining the table on page 100, *it will be noticed* (1) *that for whole numbers the characteristic is a figure which is one less in value than the number of figures in the number whose log we are considering,* (2) *that for decimals the* characteristic is a negative *number, and is one greater than the number of ciphers after the decimal point.*

Thus
log	1000	(4 figures) =	3
,,	600	(3 ,,) =	2 + a fraction
,,	9	(1 ,,) =	0 + ,,
,,	·01	(1 cipher) =	−2
,,	·026	(1 ,,) =	−2 + a fraction
,,	·0025	(2 ,,) =	−3 + ,,

The above gives us the working rule for obtaining the characteristic.

104. To obtain the mantissa.

This is found from a table of logarithms (page 238).

On examining the table of 4-figure logarithms it will be noticed that it is composed of 3 sets of columns.

COMMON LOGARITHMS

	0	1	2	3	4	5	6	7	8	9	1	2	3	4	5	6	7	8	9
50	6990	6998	7007	7016	7024	7033	7042	7050	7059	7067	1	2	3	3	4	5	6	7	8
51	7076	7084	7093	7101	7110	7118	7126	7135	7143	7152	1	2	3	3	4	5	6	7	8
52	7160	7168	7177	7185	7193	7202	7210	7218	7226	7235	1	2	2	3	4	5	6	7	7
53	7243	7251	7259	7267	7275	7284	7292	7300	7308	7316	1	2	2	3	4	5	6	6	7
54	7324	7332	7340	7348	7356	7364	7372	7380	7388	7396	1	2	2	3	4	5	6	6	7
55	7404	7412	7419	7427	7435	7443	7451	7459	7466	7474	1	2	2	3	4	5	5	6	7
56	7482	7490	7497	7505	7513	7520	7528	7536	7543	7551	1	2	2	3	4	5	5	6	7
57	7559	7566	7574	7582	7589	7597	7604	7612	7619	7627	1	2	2	3	4	5	5	6	7

Set 1. A single column containing numbers 10 to 99.
Set 2. Ten columns numbered 0 to 9.
Set 3. Nine columns numbered 1 to 9.

EXAMPLE 1. Find log of 50.

Since this number lies between 10 and 100 its log lies between 1 and 2

$$\therefore \quad \log 50 = 1 + \text{a fraction.}$$

To find this fraction, namely the mantissa, go down the first column until 50 is reached. Opposite 50, and under column headed 0 in Set 2 the figures 6990 are found. This is the fractional part required.

$$\therefore \quad \log 50 = 1 \cdot 6990.$$

EXAMPLE 2. Find log of 576.

Proceeding as before, $\log 576 = 2 + \text{a fraction}$. Go down col. 1 until 57 is reached. Then under col. 6 in Set 2, and opposite 57, the figures 7604 appear.

$$\therefore \quad \log 576 = 2 \cdot 7604.$$

EXAMPLE 3. Find log 5768.

$$\log 5768 = 3 + \text{a fraction.}$$

Proceed as in Example 2. Having found the figures 7604, under col. 6, go to the last set of columns. Go down col. 8, until opposite 5 in col. 1, and the figure 6 is seen. Add this figure 6 to the digit in the fourth decimal place of 7604 which gives us 7610.

$$\therefore \quad \log 5768 = 3 \cdot 7610.$$

EXAMPLE 4. Find log 0·05473.

$$\log 0 \cdot 05473 = \bar{2} + \text{a fraction.}$$

As in the former example 547 gives 7380, and the figure under col. 3 in the last set gives 2.

log 0·0547	$= \bar{2} \cdot 7380$
Difference corresponding to $3 =$	2
$\therefore \quad$ log 0·05473	$= \overline{2 \cdot 7382}$

Note 1. *The addition of the number from the difference column (Set 3) is usually performed mentally.*

Note 2. *The negative sign, it will be noticed, is written* over *the characteristic, as* **the mantissa part of the logarithm is positive**

Exercises 42

Find the logarithms of the following numbers:

1. 5678, 567·8, 56·78, 5·678.
2. 3·509, 0·3509, 0·03509, 0·003509.
3. 38·79, ·03879, 387900.
4. 1·452, 3·147, 0·7854.
5. 30·09, 0·0006, 156·2.

ANTILOGARITHMS

105. If you turn to the next set of tables at the end of the book, you will see that they are called Antilogarithms. By their aid, the value of 10 to any power may be obtained. The method employed is similar to that in reading the logarithm of a number.

EXAMPLE 1. Find the value of $10^{0 \cdot 1618}$.

$$10^0 = 1.$$

∴ $10^{0 \cdot 1618}$ must lie between 1 and 10.

Go down left-hand column until ·16 is reached. Opposite this under col. 0 the figures 1445 are seen.

$$∴   10^{0 \cdot 16} = 1 \cdot 445.$$

Again, opposite ·16 and under col. 1 are seen the figures 1449.

$$∴   10^{0 \cdot 161} = 1 \cdot 449.$$

Now go to col. 8 in the last set of columns and opposite ·16 the number is found. Adding this to 1·449 we get 1·452.

$$∴   10^{0 \cdot 1618} = 1 \cdot 452.$$

Similarly $10^{1 \cdot 1618} = 14 \cdot 52,$

$$10^{2 \cdot 1618} = 145 \cdot 2,$$

$$10^{3 \cdot 1618} = 1452.$$

It will be noticed that the characteristic determines the position of the decimal point in the answer.

EXAMPLE 2. Find the value of $10^{1 \cdot 7935}$.

The value must lie between 10 and 100.

$$10^{0 \cdot 79}   = 6 \cdot 166,$$

$$10^{0 \cdot 793} = 6 \cdot 209,$$

$$10^{0 \cdot 7935} = 6 \cdot 216,$$

$$10^{1 \cdot 7935} = 62 \cdot 16,$$

that is, 62·16 is the number whose log is 1·7935.

Exercises 43

Find the value of:

1. $10^{1 \cdot 0719}$, $10^{1 \cdot 0052}$, $10^{1 \cdot 1234}$.

2. $10^{2 \cdot 16}$, $10^{\bar{1} \cdot 0259}$, $10^{\bar{1} \cdot 4972}$.

3. $10^{0 \cdot 699}$, $10^{1 \cdot 699}$, $10^{\bar{1} \cdot 699}$.

Find the numbers whose logs are:

    4. 1·6916, 2·142, 0·4972.

    5. 0·3647, $\bar{1}$·0009, $\bar{2}$·5079.

106. Multiplication by means of logarithms.

Two or more quantities of the same kind are multiplied togethe
by adding their indices, e.g. $a^2 \times a^4 = a^6$. Similarly

$$10^{1\cdot26} \times 10^{1\cdot24} = 10^{2\cdot5}$$

but $10^{1\cdot26} = 18\cdot2,$

and $10^{1\cdot24} = 17\cdot38,$

and $10^{2\cdot5} = 316\cdot2.$

         $\therefore$ $18\cdot2 \times 17\cdot38 = 316\cdot2.$

**Therefore it is seen that to multiply two or more number
together, we add the logs of the numbers so to be multiplie
and this gives the log of the result.**

Example. $3\cdot782 \times 19\cdot76.$

$$\left.\begin{array}{l} \log\ 3\cdot782 = 0\cdot5777 \\ \log 19\cdot76\ = 1\cdot2958 \end{array}\right\} \text{Adding}$$

$\log 3\cdot782 + \log 19\cdot76 = 1\cdot8735 = \log 74\cdot73.$

       $\therefore$ $3\cdot782 \times 19\cdot76 = 74\cdot73.$

Steps.   (1) From tables find logs of 3·782, 19·76.

          (2) Add them.

          (3) From tables of antilogs obtain result.

Exercises 44

Evaluate by logs:

    1. $2\cdot59 \times 1\cdot361.$ **2.** $34\cdot61 \times 2\cdot481.$

    3. $3\cdot079 \times 3\cdot142.$ **4.** $0\cdot1064 \times 1\cdot57.$

    5. $20\cdot07 \times 0\cdot03569.$ **6.** $0\cdot2503 \times 0\cdot425.$

107. Division by means of logarithms.

It has already been shown (para. 48) that $x^5 \div x^3 = x^{5-3} = x^2$. Tha
is, *in division* of a quantity by another of the same kind, *the power
are subtracted*. This gives the power of the quantity in the result.

Example 1. Evaluate $10^{2\cdot16} \div 10^{1\cdot43}$.

    $10^{2\cdot16} \div 10^{1\cdot43} = 10^{2\cdot16-1\cdot43} = 10^{0\cdot73} = 5\cdot37.$

Instead of having numbers expressed to the power of 10 it is more usual to have the numbers expressed as decimals. Thus as $10^{2 \cdot 16} = 144 \cdot 5$, and $10^{1 \cdot 43} = 26 \cdot 92$, the above question could be written:

Evaluate $144 \cdot 5 \div 26 \cdot 92$.

From what has been shown it is clear that **the procedure is to subtract the log of the divisor from the log of the dividend.** This will give the log of the result, or in other words, the index of the power of 10 which may be evaluated by using the table of antilogarithms.

EXAMPLE 2. Divide $4 \cdot 623$ by $5 \cdot 051$.

$$\left.\begin{array}{l} \log 4 \cdot 623 = 0 \cdot 6649 \\ \log 5 \cdot 051 = 0 \cdot 7033 \end{array}\right\} \text{Subtract.}$$

Here a difficulty arises because a greater number has to be subtracted from a lesser. This difficulty is overcome by writing $0 \cdot 6649$ as $1 \cdot 6649 - 1$. The number $0 \cdot 7033$ may now be subtracted from $1 \cdot 6649$. Thus

$$1 \cdot 6649 - 0 \cdot 7033 - 1 = \cdot 9616 - 1.$$

The minus 1 is now placed in front and written $\bar{1} \cdot 9616$, and this is the log of the answer. Turning to the table of antilogs, $0 \cdot 9616$ gives the result 9154, and since the characteristic of the result is $\bar{1}$, the result must lie between $0 \cdot 1$ and 1. Thus the result is $0 \cdot 9154$.

EXAMPLE 3. Divide $0 \cdot 6255$ by $0 \cdot 03142$.

$$\log 0 \cdot 6255 = \bar{1} \cdot 7962,$$
$$\log 0 \cdot 03142 = \bar{2} \cdot 4972.$$

The seeming difficulty of subtracting $\bar{2} \cdot 4972$ from $\bar{1} \cdot 7962$ will disappear if the following method is understood:

$$\bar{1} \cdot 7962 = 0 \cdot 7962 - 1, \ \bar{2} \cdot 4972 = 0 \cdot 4972 - 2.$$

$$\begin{aligned} \therefore \quad \bar{1} \cdot 7962 - \bar{2} \cdot 4972 &= (0 \cdot 7962 - 1) - (0 \cdot 4972 - 2) \\ &= 0 \cdot 7962 - 1 - 0 \cdot 4972 + 2 \\ &= 0 \cdot 7962 - 0 \cdot 4972 + 1 \\ &= 1 \cdot 7962 - 0 \cdot 4972 \\ &= 1 \cdot 299. \end{aligned}$$

This is the log of the answer, that is, the answer is $10^{1 \cdot 299}$. Reading from the table of antilogarithms, $1 \cdot 299$ gives 1991, and since the answer is $10^{1 \cdot 299}$ it must lie between 10 and 100. It is therefore $19 \cdot 91$, or in other words, since 1 is the characteristic, there are two figures in front of the decimal point in the answer.

108. From the foregoing the following important rule may be deduced.

A minus sign in front of a negative characteristic changes it into a positive characteristic.

Study the following:

$$(1) \quad 4 + \bar{1} = 4 + (-1) = 4 - 1 = 3.$$
$$(2) \quad 3 + \bar{4} = 3 + (-4) = 3 - 4 = -1 = \bar{1}.$$
$$(3) \quad \bar{3} + \bar{4} = -3 + (-4) = -7 = \bar{7}.$$
$$(4) \quad \bar{1} - \bar{2} = -1 - (-2) = -1 + 2 = 1.$$
$$(5) \quad 0 - \bar{3} = 3.$$
$$(6) \quad 0 - 3 = 0 - (+3) = \bar{3}.$$

Thus in Example 3 (para. 107) we have:

$$\text{From } \log 0 \cdot 6255 \ = \bar{1} \cdot 7962$$
$$\text{Take } \ \log 0 \cdot 03142 = \bar{2} \cdot 4972$$
$$\overline{ 1 \cdot 2990}$$

The positive decimal parts are subtracted in the ordinary way. For characteristic we have $\bar{1} - \bar{2} = -1 - (-2) = -1 + 2 = 1$.

Exercises 45

Evaluate by logarithms:

1. $17 \cdot 42 \div 3 \cdot 56.$ **2.** $216 \cdot 8 \div 72 \cdot 31.$

3. $2 \cdot 413 \div 12 \cdot 27.$ **4.** $0 \cdot 4567 \div 0 \cdot 03709.$

5. $0 \cdot 0431 \div 0 \cdot 512.$ **6.** $3 \cdot 719 \div 6 \cdot 725.$

109. Evaluation of roots and powers by logarithms.

Since $\qquad\qquad 9^2 = 9 \times 9.$

$$\therefore \quad \log 9^2 = \log (9 \times 9)$$
$$= \log 9 + \log 9$$
$$= 2 \log 9.$$

Similarly it may be shown that

$$\log 8^3 = 3 \log 8.$$
$$\therefore \quad \log a^p = p \log a.$$

This is true for all values of p.

EXAMPLE 1. Evaluate $(0.3659)^2$.

$$\log (0.3659)^2 = 2 \log 0.3659 = 2 \times \bar{1}.5634$$
$$= 2 \,(-1 + 0.5634)$$
$$= -2 + 1.1268$$
$$= -2 + 1 + 0.1268$$
$$= -1 + 0.1268 = \bar{1}.1268.$$
$$\therefore \quad 10^{\bar{1}.1268} = 0.1339.$$

EXAMPLE 2. Evaluate $(1.5236)^{\frac{1}{3}}$.

$$\log (1.5236)^{\frac{1}{3}} = \tfrac{1}{3} \log 1.5236$$
$$= \tfrac{1}{3} \times 0.1827$$
$$= 0.0609.$$
$$\therefore \quad 10^{0.0609} = 1.15.$$

EXAMPLE 3. Evaluate $(0.0056)^{\frac{1}{2}}$.

$$\log (0.0056)^{\frac{1}{2}} = \tfrac{1}{2} \log 0.0056$$
$$= \tfrac{1}{2} \times \bar{3}.7482.$$

In this case if -3 is divided by 2 there is a remainder of -1, which evidently cannot be combined with the following 7 to make 17, as one is negative and the other positive. In cases like this, the following is the procedure.

When the negative characteristic is not exactly divisible, make it so, by adding to it a negative quantity, and at the same time add an equal positive quantity to the mantissa.

In Example 3, as the characteristic -3 is not exactly divisible by 2, make it so by adding -1 to it, and at the same time add $+1$ to the mantissa. This in effect converts $\bar{3}.7482$ into $\bar{4} + 1.7482$. Dividing now by 2 we get $\bar{2}.8741$ which is the log of the answer.

$$\therefore \quad 10^{\bar{2}.8741} = 0.07484.$$

EXAMPLE 4. Evaluate $(0.8651)^{-2}$.

$$\log (0.8651)^{-2} = -2 \log 0.8651 \quad \big| \quad \bar{1}.9371$$
$$= -2 \times \bar{1}.9371 \quad \big| \quad = \bar{1} + .9371$$
$$= -2 \times -0.0629 \quad \big| \quad = -0.0629$$
$$= 0.1258$$
$$\therefore \quad 10^{0.1258} = 1.336.$$

EXAMPLE 5. Evaluate $\dfrac{0.3652}{0.0465} + \dfrac{1}{(2.69)^2}$.

The value of each of the terms must be found separately for the operations of addition and subtraction cannot be performed logarithmically.

$$\log 0.3652 = \bar{1}.5625$$
$$\log 0.0465 = \bar{2}.6675$$
$$\log \text{quotient} = \overline{0.8950}$$

$$\log \frac{1}{(2.69)^2} = \log 1 - 2 \log 2.69$$
$$= 0 - 2 \times 0.4298$$
$$= 0 - .8596$$
$$= \bar{1}.1404.$$

$$\therefore \quad \frac{0.3652}{0.0465} = 7.852$$

$$\therefore \quad \frac{1}{(2.69)^2} = 0.1381.$$

$$\therefore \quad \frac{0.3652}{0.0465} + \frac{1}{(2.69)^2} = 7.852 + 0.1381 = 7.9901.$$

Summary.

1. *Add* logarithms for *multiplication*.
2. *Subtract* logarithms for *division*.
3. *Multiply* logarithms by 2, 3, etc., to find 2nd, 3rd, etc., *powers*.
4. *Divide* logarithms by 2, 3, etc., to find 2nd, 3rd, etc., *roots*.

Exercises 46

Find the values of:

1. $(82.64)^{\frac{2}{3}}$.

2. $0.00081 \times (21)^3 \times 730$.

3. $\sqrt{(6.231)^3}$.

4. $\dfrac{20.51 \times \sqrt{4.86}}{(5.31)^2}$.

5. $\left(\dfrac{129.8}{3.142 \times 7.96}\right)^{\frac{1}{4}}$.

6. $\dfrac{0.6745\,(1 - r^2)}{\sqrt{376}}$; $r = 0.675$.

NOTE. $(1 - r^2) = (1 - r)\,(1 + r)$.

7. $(0.1628)^{-2.7}$.

8. $\dfrac{(4.273)^{\frac{2}{3}}}{(0.04273)^{\frac{1}{3}}}$.

9. $\sqrt[4]{27.31} \times \sqrt[3]{0.374} \div (0.837)^3$.

10. $\dfrac{\sqrt[3]{4287 \times 0.0023}}{0.124 \times 3.729}$.

11. $T = \dfrac{\pi}{16} d^3 f$. If $T = 10{,}260$, $f = 6000$, find d.

12. The time in seconds to send a packet by pneumatic transmission through a tube is given by

$$t = 0.000482 \sqrt{\frac{l^3}{Pd}},$$

where l is the length, d the diameter, and P the pressure. Find t, when $l = 2900$, $d = \frac{3}{16}$, $P = 10$.

13. The formula $N = \dfrac{c^2 u^2}{d^2 (u-v)^2}$ occurs in connection with experiments on the velocity of sound. Find N where $c = 1 \cdot 61$, $d = 10 \cdot 8$, $u = 334 \cdot 2$, $v = 330 \cdot 6$.

14. The area of a triangle is given by
$$A = \sqrt{s\,(s-a)\,(s-b)\,(s-c)}.$$
If $a = 4 \cdot 43$, $b = 4 \cdot 1$, $c = 3 \cdot 64$, $s = \frac{1}{2}\,(a+b+c)$, find A.

15. The strength C in amperes of an alternating current in a circuit is given by
$$C = \frac{E}{\sqrt{R^2 + n^2 L^2}}.$$
Find C when $E = 106$ volts, $R = 2 \cdot 4$ ohms, $n = 54$, $L = 0 \cdot 017$.

16. If $T = 2\pi \sqrt{\dfrac{h^2 + k^2}{hg}}$, find k when $T = 11$, $g = 32$, $h = 4$, $\pi = 3\frac{1}{7}$.

17. The length l of a brass wire under a pull of W lb. is given by the equation
$$l = 10 + \frac{W}{11,500}.$$
What pull will stretch it to a length of $10 \cdot 13$ in.?

18. The approximate diameter of wire (in inches) to carry a given current C with $\theta°$ rise in temperature can be obtained from,
$$D = \sqrt[3]{\frac{4JpC^2}{\pi^2 m\theta}}.$$
Find D when $J = 0 \cdot 0935$, $\theta = 35$, $p = \dfrac{1 \cdot 6}{10^6}$, $C = 55$, $m = 0 \cdot 0025$ and $\pi = 3 \cdot 142$.

19. Evaluate $\dfrac{(3 \cdot 56)^2 + (0 \cdot 945)^{\frac{1}{2}}}{\log 13 \cdot 72}$.

20. Find the index of the power to which 10 must be raised to give a result of 16. Given $\log 2 = 0 \cdot 3010$.

CHAPTER X

CIRCUMFERENCE OF A CIRCLE, Fig. 65 (a)–(e)

110. (a) The circumference of a circle may be measured by wrapping a strip of paper round a cylinder and pricking through the overlapping ends. The distance between the pin-holes, i.e. PP_1 in the diagram, is the circumference. If the length of the circumference be

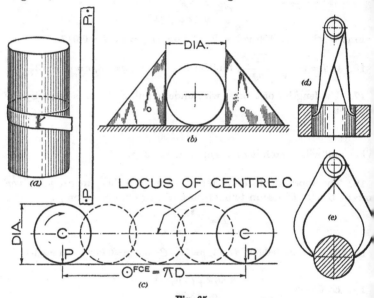

Fig. 65

divided by the diameter, the ratio $\dfrac{\text{circumference}}{\text{diameter}}$ is found, in any case, to approximate to 3·14. We employ the Greek letter π (pronounced *pi*) to denote this ratio.

(b), (d), (e). Methods of measuring diameters are shown at (b), (d), (e). At (b) we employ set-squares, at (d) inside calipers, and at (e) outside calipers.

(c) Here is shown a circular disc having a point P marked on it rcumference. Suppose the disc to roll for one revolution along a raight line. Then PP_1 gives the length of the circumference of the role.

Summarising.

$$\frac{\textbf{Circumference}}{\textbf{Diameter}} = \pi.$$

$$\therefore \quad \textbf{Circumference} = \pi D = 2\pi R$$

where D=diameter, R=radius).

The value of π is indeterminate and we have, therefore, to use approximations. For some purposes 3·14 is sufficiently accurate; sometimes it more convenient to use $\frac{22}{7}$. Where greater accuracy is essential we assume $\pi = 3\cdot1416$.

WORKED EXAMPLES.

1. A bicycle wheel is 28 in. diameter. How far does the bicycle travel or each turn of the wheel? Take $\pi = \frac{22}{7}$.

Distance travelled = circumference of wheel

$$= \pi \times \text{diameter}$$
$$= \frac{22}{7} \times 28$$
$$= 7 \text{ ft. } 4 \text{ in.}$$

2. The circumference of a flywheel was measured by a thin steel tape and found to be $39\frac{1}{4}$ in. What is its diameter? Take $\pi = 3\cdot14$.

$$d = \frac{\text{circumference}}{\pi}$$
$$= \frac{39\cdot25}{3\cdot14}$$
$$= 12\cdot5 \text{ in.} = 1 \text{ ft. } 0\frac{1}{2} \text{ in.}$$

Exercises 47

1. A wheel is 3·45 ft. in diameter and makes 510 revolutions in rolling ong a road. What distance in feet is passed over? ($\pi = 3\cdot14$.)

2. Find, to the nearest inch, the length of the circumference of a comotive driving wheel if it is 7 ft. diameter. How many revolutions ill the wheel make whilst the engine moves $\frac{1}{2}$ mile? ($\pi = \frac{22}{7}$.)

3. Fig. 66 shows an arrangement of parallel shafts and equal pulleys. What length of belting will be required to drive one shaft from the other? $= \frac{22}{7}$.)

4. Find, to the nearest foot, the length of rope 1 in. thick required t
make 15 turns round a drum or cylinder 2 ft. 3 in. diameter. ($\pi = \frac{22}{7}$.

(*Hint. In this and similar cases it is necessary to work to the mea
diameter of the rope as it lies round the drum. See the centre-line circle i
Fig. 67. Its diameter* = 2 ft. 4 in.)

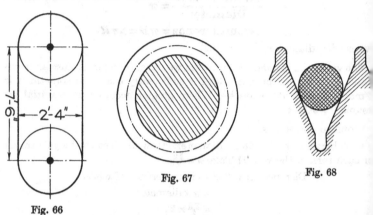

2'-4"

Fig. 67

Fig. 68

Fig. 66

5. Hempen ropes of circular section are largely used for power trans
mission. See Fig. 68. In the trade they are measured by their circum
ferences. Hence a 6 in. rope is of 6 in. circumference,
or about $1\frac{7}{8}$ in. diameter. State the diameters of the
following ropes: (*a*) 4 in., (*b*) 5 in., (*c*) 7 in., (*d*) 9 in.
(Answers to nearest sixteenth of an inch.)

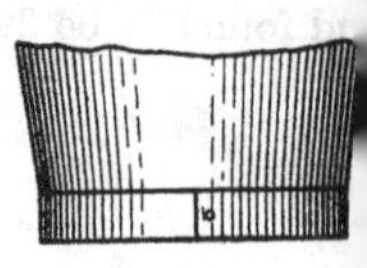

6. What length of hoop iron is required to form
the feet of 2 gross of buckets (Fig. 69), the bottom

Fig. 69

diameter of each being 9 in., and the overlap for the riveting of eac
hoop being 1 in.? ($\pi = \frac{22}{7}$.)

111. Gear wheel teeth–pitch.

Spur wheels are toothed discs, i.e. cylinders having teeth cut acros
their rims. They are used to transmit motion between *parallel shaft
lying in the same plane.*

How the teeth are measured. Fig. 70 shows very clearly the section o
the rim of a typical wheel. There are three concentric circular arcs o
which the pitch circle is the most important for along it, and from it

any measurements are taken. For instance **circular pitch** is shown
the distance from a point on one tooth to a corresponding point on the
next *measured round the pitch circle.*

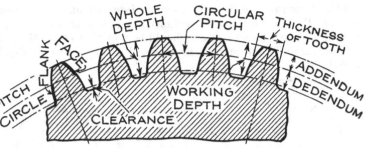

Fig. 70

Let
$$d = \text{diameter of pitch circle,}$$
$$CP = \text{circular pitch,}$$
$$n = \text{no. of teeth in wheel.}$$

Then   Circumference of pitch circle $= CP \times n$.

But   ,,   ,,   ,, $= \pi d$.

$$\therefore \quad CP \times n = \pi d.$$

$$\therefore \quad CP = \frac{\pi d}{n}; \quad n = \frac{\pi d}{CP}; \quad d = \frac{CP \times n}{\pi}.$$

Exercises 48

$$(\pi = \tfrac{22}{7})$$

1. A spur wheel has 20 teeth of $\frac{3}{8}$ in. CP. Find the diameter of the
pitch circle.

2. Find the pitch circle diameter of a spur wheel of $1\frac{1}{4}$ in. CP if it
has 30 teeth.

3. The pitch circle diameter of a spur wheel is 11·419 in. It has
11 teeth. Find the pitch of the teeth.

4. A spur wheel of 46 teeth has a pitch circle diameter of 20·133 in.
Find the circular pitch of the teeth.

5. A bevel wheel has 24 teeth of $1\frac{1}{8}$ in. CP. Find the diameter of the
pitch circle.

6. The pitch diameter of a wheel is 17·19 in.; its $CP = 1\frac{1}{2}$ in. Find
the number of teeth.

THE AREA OF A CIRCLE

112. Describe a circle and divide it into any number of equal sectors. For the purpose of clear illustration 16 sectors are taken in Fig. 71, but on a larger figure it should be possible to take 24. These sectors, carefully cut out, can be arranged side by side as in Fig. 71 *ABCD*.

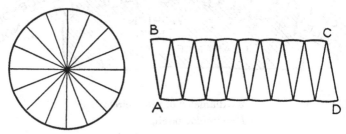

Fig. 71

The resulting figure clearly approximates to a parallelogram—indeed the resemblance would be far closer had we divided the circle into a greater number of sectors. Examine the approximate parallelogram, and note that:

(1) its base is approximately equal to half the circumference of the circle;

(2) its vertical height is approximately equal to the radius of the circle;

(3) its area $= \frac{1}{2}$ circumference × radius

$$= \frac{1}{2} \times 2\pi r \times r$$
$$= \pi r^2.$$

Note. $r = \dfrac{d}{2}, \quad \therefore \quad r^2 = \dfrac{d^2}{4}.$

$$\therefore \quad \pi r^2 = \frac{\pi d^2}{4} = 0 \cdot 7854 d^2.$$

Conclusion.

Area of circle $= \pi r^2 = \dfrac{\pi}{4} d^2 = 0 \cdot 7854 d^2.$

To find radius or diameter when the area is known, we have:

$$r^2 = \frac{A}{\pi}, \qquad \therefore \quad r = \sqrt{\frac{A}{\pi}}.$$

$$d^2 = \frac{4A}{\pi}, \qquad \therefore \quad d = \sqrt{\frac{4A}{\pi}}.$$

EXAMPLE 1. Find the area of a circle 10 in. diameter.

Using both radius and diameter formulae, and taking $\pi = 3 \cdot 14$:

(a) Area $= \pi r^2 = 3 \cdot 14 \times 5^2 = 3 \cdot 14 \times 25 = 78 \cdot 5$ sq. in.;

(b) Area $= \frac{\pi}{4} d^2 = \frac{3 \cdot 14 \times 100}{4} = 78 \cdot 5$ sq. in.

EXAMPLE 2. The area of a circle is $43 \cdot 24$ sq. in. What is its diameter?

$$\text{Area} = \frac{\pi}{4} d^2 = 43 \cdot 24,$$
$$\therefore \quad \pi d^2 = 4 \times 43 \cdot 24,$$
$$\therefore \quad d^2 = \frac{4 \times 43 \cdot 24}{\pi} = \frac{172 \cdot 96}{\pi}$$
$$= 55 \cdot 0825,$$
$$\therefore \quad d = \sqrt{55 \cdot 0825} = 7 \cdot 42 \text{ in.}$$

Exercises 49

$$(\pi = \tfrac{22}{7})$$

1. Fig. 72 shows the section of a short tunnel. Find the area of its cross-section in square feet and square yards.

2. A circular disc is 3 ft. in diameter and is subjected to atmospheric pressure. If the total pressure on the disc is 136 cwt., give the intensity of pressure in lb. per sq. in.

3. A wall containing a circular window 3 ft. in diameter and a door 6 ft. 6 in. by 3 ft. is 14 ft. long by 11 ft. high. Find, in square feet, the area to be plastered.

4. Find the diameters of circles having the following areas (a) 100 sq. in., (b) 1386 sq. ft.

5. If the circumference of a circle is 880 yd. in length, find its area in square yards.

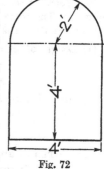

Fig. 72

6. If the area of a semi-circle is 2772 sq. in., what is the area of its circumscribing rectangle?

7. Draw a square $ABDC$ having its sides 4 in. long. With D as centre draw the quadrant $BQCD$, and with A as centre draw the quadrant $BPCA$. Find the area of the space enclosed by the two quadrantal arcs BPC and BQC (Fig. 73).   (*C. and G.*)

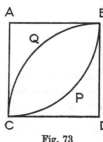

8. A ceiling is circular and 21 ft. in diameter. It cost £4. 4s. 0d. to plaster. How much would it cost to plaster a rectangular ceiling 30 ft. by 20 ft. ?

113. Area of a sector.

1. A sector having an angle of 1° between its radii will evidently have an area $\frac{1}{360}$ that of the circle. Thus if the angle between the radii is $n°$,

Fig. 73

$$\text{Area of sector} = \frac{n}{360} \times \text{area of circle.}$$

2. Similarly length of arc of sector (see AB in Fig. 74)

$$= \frac{n}{360} \times \text{circumference of circle.}$$

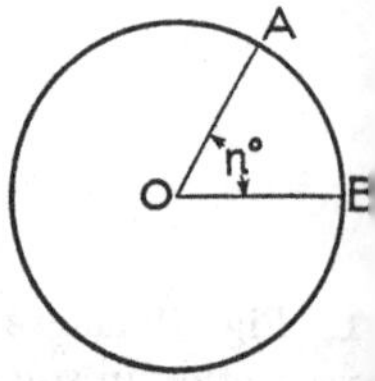

3. We saw in para. 112 that the area of a circle is equal to half the circumference multiplied by the radius. Similarly **the area of a sector equals half the arc multiplied by the radius.**
Thus in Fig. 74 area of sector $AOB = \frac{1}{2}AB \times OB$.

EXAMPLE. A sector is cut from a circle of 5 in. diameter. The angle between the radii is 60°. Find
(1) the length of the arc of the sector, (2) the area of the sector.

Fig. 74

(1) Length of arc $= \frac{60}{360} \times$ circumference of circle
$\qquad\qquad = \frac{1}{6} \times \pi d = \frac{1}{6} \times 15\cdot708$
$\qquad\qquad = 2\cdot618$ in.

(2) Area of sector $= \frac{1}{2}$ (arc $\times$ radius)
$\qquad\qquad = \frac{1}{2} \times 2\cdot618 \times 2\frac{1}{2}$
$\qquad\qquad = 3\cdot27$ sq. in. $+$

114. Area of an annulus.

An annulus is the plane figure contained between two concentric circles, i.e. circles described from the same centre (Fig. 75).
The area of an annulus is readily found by subtracting the area o

he smaller circle from that of the larger. This operation is simplified
y the use of factors (difference between two squares).

Let R = radius of larger circle,

           r = radius of smaller circle.

Area of annulus

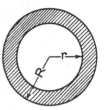

$$= \pi R^2 - \pi r^2$$
$$= \pi (R^2 - r^2)$$
$$= \pi (R+r)(R-r) \qquad \ldots\ldots(1)$$

r, alternatively,

$$= \frac{\pi}{4}(D+d)(D-d). \qquad \ldots\ldots(2)$$

Fig. 75

EXAMPLE. Find the area, to 3 significant figures, of a flat ring whose
imensions are, outside diameter = $6\frac{1}{2}$ in., inside diameter = $4\frac{1}{2}$ in.

$$\text{Area} = 0 \cdot 7854 \left(6\tfrac{1}{2} + 4\tfrac{1}{2}\right)\left(6\tfrac{1}{2} - 4\tfrac{1}{2}\right)$$
$$= 0 \cdot 7854 (11)(2)$$
$$= 17 \cdot 2788 = 17 \cdot 3 \text{ sq. in. (to 3 significant figures)}.$$

Exercises 50
$$(\pi = \tfrac{22}{7})$$

1. Fig. 76 shows the outline plan of a segmental bay window. Cal-
ulate the length of the curved portion of the faced
rickwork. ($\pi = \frac{22}{7}$.)

2. A sector having an angle of 40° between its
adii is cut from a circle of 10 in. diameter. Find
) the length of the arc of the sector, (b) the area
f the sector.

Fig. 76

3. A circular arc is 4·817 in. long; the radius is 12 in. What is the angle
degrees subtended by the arc at the centre?

4. The foundations of a circular tower are in the form of an annulus
aving an outside diameter of 26 ft., and an inside diameter of 14 ft.
ind the cost of damp-proofing at 1s. 4d. per square foot.

5. A circular flagged path has its outside diameter 12 ft. 6 in. and its
nside diameter 9 ft. 3 in. Find the cost of paving it at 1s. 0½d. per
quare foot (to nearest penny).

6. A marine engine propeller shaft is hollow and has an outside
iameter of 18 in. and an inside diameter 8¼ in. Find its cross-sectional
rea to the nearest square inch.

7. Calculate, in due order from left to right, the areas of the shaded portions of each diagram in Fig. 77.

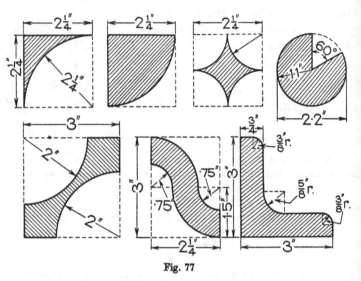

Fig. 77

8. (*a*) Find the area of a circular racing track, outside diameter 2000 ft., inside diameter 1900 ft.

(*b*) How much would it cost to enclose it, both inside and outside, by circular fencing at 2*s.* per yard?

CHAPTER XI

USE OF SQUARED PAPER

115. The relationship between two quantities which depend upon each other for their values may be shown with advantage by means of a **graph**.

EXAMPLE 1. The working expenses of the Great Western Railway Company from 1913 to 1922 are represented by the following figures.

Year	1913	1914	1915	1916	1917	1918	1919	1920	1921	1922
Cost	65	65	67	66	70	73	79	84	83	81

Fig. 78 shows the information in graphic form. On squared paper are drawn two straight lines, one horizontal and one vertical. We call these lines OX and OY respectively.

On the horizontal line (OX) mark off equal distances, commencing at the point O, to represent years, placing the year 1913 at O.

On the vertical line (OY), from the same starting point, mark off equal distances to represent units of expenditure.

In the year 1913, the expenses are represented by 65, therefore the point 65 is marked on the vertical line opposite 1913. Go next to 1914. In that year 65 again represents expenses, so on the vertical line, opposite 1914, 65 is marked as shown. On each vertical line opposite each succeeding year, a point is marked representing the expenses for that year.

The points are now joined up by straight lines. The resulting figure is a graphic representation of the expenditure between 1913 and 1922, and conveys a much clearer idea than any tabulation of figures could.

It shows at a glance:

(1) There was no increase in expenses between 1913 and 1914.

(2) There was the greatest rate of increase in expenses between 1918 and 1919.

(3) There was the greatest rate of decrease in expenses between 1921 and 1922.

This process is called *plotting* the information given.

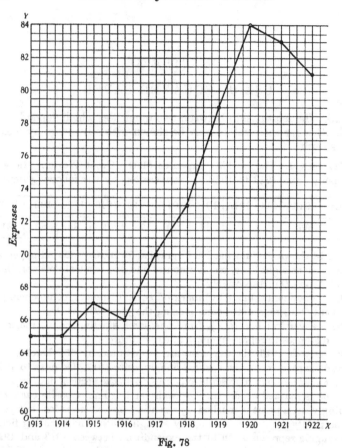

Fig. 78

EXAMPLE 2. The speed of a non-stop train between London and Cheltenham was observed to vary as follows. Plot a graph.

Miles from London	10	20	30	40	50	60	70	80	90	100
Speed m.p.h.	30	50	65	70	85	88	80	75	45	25

On squared paper draw two lines, one horizontal and one vertical as in Fig. 79.

On the horizontal line mark off lengths to represent distance in miles
from London and on the vertical mark off lengths to represent speed, in
miles per hour.

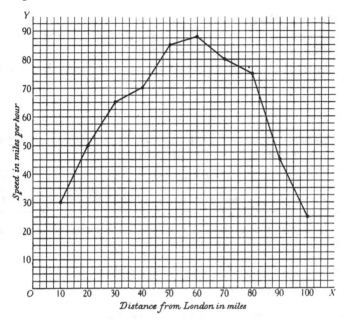

Fig. 79

When the train is 10 miles from London its speed is 30 m.p.h. Therefore opposite 10 on the horizontal line the point 30 is marked on the
vertical line; opposite 20 on the horizontal line the point is marked
opposite 50 on the vertical line. The remaining points are plotted
similarly. These points are then joined up as shown in Fig. 79 and a
graphic representation of the data is given. From it can be seen at a
glance at what stage of the journey the train was gathering greatest
speed, and between what two distances the rate of travelling was
lessening most rapidly.

From the graph may also be read (say):

(1) What the rate of travelling was in m.p.h. when the train was
5 miles from London.

(2) The distances it was from London when travelling at 60 m.p.h.

This reading of intermediate values from a graph is called **Interpolation**.

Hints on plotting.

1. Always make your scales as large as your paper will allow Small scales are liable to give poor results.

2. The readings on the horizontal and vertical lines (called **Axes** need not commence at zero.

Exercises 51

1. The following table gives the maximum temperature for the firs fortnight in August at Kew. Plot the graph.

Aug.	1	2	3	4	5	6	7	8	9	10	11	12	13	14
Temp.	66	67	70	65	61	68	66	66	65	65	65	60	62	63

2. The figures representing cost of living in a certain year were a follows:

Month	Jan.	Feb.	Mar.	Apr.	May	June	July	Aug.	Sept.	Oct.
Cost	78	77	76	74	70	69	69	71	73	75

Draw a graph showing rise and fall in cost of living.

3. The tonnage of ships using oil fuel, is given in the following tabl Draw a graph, and state between which two years there was the mos rapid increase.

Year	1919	1920	1921	1922	1923
Millions of tons	$5\frac{1}{4}$	$9\frac{1}{4}$	$12\frac{3}{4}$	$14\frac{1}{2}$	$15\frac{3}{4}$

4. The following table shows the highest temperatures observed a Greenwich between the years 1914 and 1923. Construct a graph.

Year	1914	1915	1916	1917	1918	1919	1920	1921	1922	1923
Temp.	92·1	87·2	83·9	93·2	89·8	87·5	86·5	94	90·6	92·2

5. The following table shows the lowest temperatures observed over the same period. Construct a graph.

Temp.	19·9	22·3	23·4	17·2	18·5	15·5	15·7	25·3	24·4	24·4

6. A tradesman's receipts for a week are as follows. Show by a graph.

Day	Mon.	Tues.	Wed.	Thur.	Fri.	Sat.
Receipts	£2. 10s.	£8. 15s.	£6. 5s.	£5. 15s.	£7	£12. 15s.

7. The following was the rainfall in inches at a certain place. Plot a graph.

Month	Jan.	Feb.	Mar.	Apr.	May	June
Rainfall	2·4	2·9	3·2	3·4	2·5	2·0
Month	July	Aug.	Sept.	Oct.	Nov.	Dec.
Rainfall	1·8	0·7	1·5	1·9	3·6	2·9

8. The annual premium required by a certain life assurance society to secure £100 at death is given by the following table. Show this graphically.

Age next birthday	15	20	25	30	35	40	45
Premium	£1. 13s.	£1. 18s.	£2. 8s.	£2. 19s.	£3. 16s.	£4. 19s.	£6. 18s.

PROPORTION BY GRAPHS

116. In addition to illustrating certain data on squared paper, graphs may be applied to the solution of problems. Many problems may be more clearly and quickly solved by a graph than by any other method.

EXAMPLE. Two dozen spanners cost 6s., find the cost of 19 spanners.

As before draw the OX and the OY axes. On one axis spanners will be represented and on the other the cost. It does not really matter what denomination is chosen for a particular axis, but the quantity that makes a change in its value, because the quantity to which it is related makes a change, is generally represented on the OY axis. In this case we shall consider that the cost of the spanners depends upon the number bought. Therefore the OY axis will represent money and the OX axis will represent spanners.

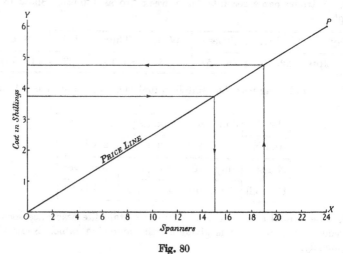

Fig. 80

(1) On the OX axis make 1 in. represent 2 spanners, and on the OY axis, let $\frac{1}{10}$ in. represent a penny.

(2) Mark off a distance, viz. 12 in., to represent 24 spanners.

(3) From this point, go vertically until opposite the point marked 6s. on the OY axis. Mark on your paper the point P.

(4) Join OP, this is called the **price line** and from it the price of any number of spanners up to 2 dozen may be read off, or the number of spanners that may be bought up to an outlay of 6s.—another example of **interpolation**.

To find the cost of 19 spanners. From point 19 on OX axis draw vertical line to *price line* and from the intersection of this line and the *price line* draw a horizontal to the OY axis. The point now arrived at is the cost of 19 spanners, viz. 4s. 9d.

To find the number that may be bought for 3s. 9d. just reverse the process.

From the point representing 3s. 9d. on the OY axis draw a horizontal line until it meets the price line; from this point draw a vertical line, until it meets the OX axis which it does at point 15. Therefore 15 spanners can be bought for 3s. 9d.

117. Distance and time.

Problems involving distance and time may be solved in the same way as the foregoing.

Hint. It is usual to represent *distance* on the OX axis and *time* on the OY axis.

Exercises 52

The following are to be solved graphically:

1. If a 20 in. length of a certain pipe weighs 30 lb., find the weight of 17 in. length of the same pipe. What length of the pipe will weigh 6 lb.?

2. A man walks at the rate of 4 miles per hour. How long will it take him to walk 19 miles? How far from home will he be in $3\frac{1}{4}$ hours?

3. The maker of cast-iron forges advertises his prices as follows:

Area of hearth plate in square feet	7·2	4·3	2·2
Price in £'s	12·3	9·3	4·8

Find the price of a hearth plate, area being 6 sq. ft.

4. The price of 15 lb. of a commodity is 3s. 9d. Find the cost of 1 lb.

5. A man walks at the rate of 3 m.p.h. He leaves home at 2 p.m. Draw a graph showing where he will be at 3.45 p.m.

6. A pole 25 ft. long casts a shadow of 40 ft. What is the height of a pole whose shadow is 24 ft.?

7. A man walks at the rate of 3 m.p.h. and a boy cycles at the rate of 15 m.p.h. They start together. Draw on the same axes their rate lines. From the point on the OX axis which marks 1 hour draw a vertical line, cutting both their rate lines. What does that part of the perpendicular, intercepted between the rate lines, represent?

8. The breaking loads on steel ropes of various girths are shown in the following table:

Girth in inches	1	2	3	4
Breaking load in tons	3·3	13	28·5	51·5

What will the probable breaking load be when the girth is 2·7 in.?

9. Plot a graph to show the square roots of the following numbers 9, 16, 25, 36, 49, 81. Use the graph to find $\sqrt{45}$, $\sqrt{60}$, $\sqrt{70}$.

CHAPTER XII

SIMULTANEOUS EQUATIONS

118. We have seen that a *simple equation* is one in which the value f one unknown quantity has to be found. We now come to con- der the problem of finding the values of *two unknown* quantities.

Supposing we know that the values of two unknown quantities x nd y are such that

$$2x + y = 7,$$

nd are asked to find the value of x and y. Several values may be und which will satisfy the equation.

If $x = 3$ and $y = 1$, then $2x + y = 7$,
r if $x = 2$,, $y = 3$, ,, $2x + y = 7$,
r again if $x = \frac{1}{2}$,, $y = 6$, ,, $2x + y = 7$.

It can therefore be seen that if only one equation is given con- ecting x and y, it is impossible to determine their values. But if we re given another equation connecting x and y then we can deter- ine what will be the *only* value for x and y which will satisfy both quations.

EXAMPLE 1. Solve: (1) $3x + y = 17$
                      (2) $5x - 2y = 21$.

Method 1. Expressing y in terms of x in Equation (1):

$$y = 17 - 3x.$$

Substitute for y in Equation (2):

$$5x - 2(17 - 3x) = 21,$$
$$\therefore \quad 5x - 34 + 6x \quad = 21,$$
$$\therefore \quad 5x + 6x \quad = 21 + 34,$$
$$\therefore \quad 11x \quad = 55,$$
$$\therefore \quad x \quad = 5,$$
d $$y \quad = 17 - 3x \quad \text{(Equation (1))}$$
$$= 17 - 15$$
$$= 2.$$

Ans. $x = 5$, $y = 2$.

Method 2. The second method of solving is by expressing y in terms of x in both equations, and then equating these two expressions and solving for x.

Take Example 1:   (1) $3x + y = 17$;

                            (2) $5x - 2y = 21$.

From Equation (1) $y = 17 - 3x$.

From Equation (2) $y = \dfrac{5x - 21}{2}$.

$$\therefore \quad \frac{5x - 21}{2} = 17 - 3x.$$

$$\therefore \quad 5x - 21 = 34 - 6x.$$

$$\therefore \quad 11x = 55,$$

whence $x = 5$.

Substituting for x in Equation (1)

$$y = 17 - 15,$$

whence $y = 2$.

Ans. $x = 5$, $y = 2$.

A third method will be dealt with later on.

EXAMPLE 2. Solve: (1) $4x + 3y = 24$;

                          (2) $5x + 2y = 23$.

Method 1. From Equation (1):

$$3y = 24 - 4x.$$

$$\therefore \quad y = \frac{24 - 4x}{3} .$$

Substitute in Equation (2) for y:

$$5x + \frac{2(24 - 4x)}{3} = 23.$$

$$\therefore \quad 15x + 48 - 8x = 69,$$

$$15x - 8x = 69 - 48,$$

$$7x = 21,$$

$$x = 3,$$

$$y = \frac{24 - 4x}{3}$$

$$= \frac{24 - 12}{3}$$

$$= 4.$$

Ans. $x = 3$, $y = 4$.

Method 2. From Equation (1): $y = \dfrac{24 - 4x}{3}$.

From Equation (2): $y = \dfrac{23 - 5x}{2}$.

$$\therefore \quad \frac{24 - 4x}{3} = \frac{23 - 5x}{2}.$$

$$\therefore \quad 2(24 - 4x) = 3(23 - 5x).$$

$$48 - 8x = 69 - 15x.$$

Transposing
$$15x - 8x = 69 - 48,$$
$$7x = 21,$$
$$x = 3,$$
hence
$$y = 4.$$

Ans. $x = 3$, $y = 4$.

Exercises 53

Solve the following equations:

1. $y = 12 - 2x$;
$6y = 5x + 4$.

2. $5x + 3y = 23$;
$2x - 4y = 4$.

3. $3a + 2b = 11$;
$4b = a + 8$.

4. $x + 12y = 54$;
$\dfrac{x}{8} + 3y = 12\tfrac{3}{4}$.

5. $3a + b = 42$;
$4a - 4b = 8$.

6. $5p = 3r$;
$4r - 5p = 5$.

7. $3x + \dfrac{5y}{2} = 8$;
$12y - 5x = 19$.

8. $\dfrac{8x}{3} + \dfrac{5y}{6} = 3$;
$6x = y + 1$.

9. $x - 4y = -4$;
$\dfrac{x}{2} + 4y = 10$.

10. $x = \dfrac{y}{4} - 1$;
$x - \tfrac{1}{5} = \dfrac{y}{5}$.

11. $\dfrac{3x}{4} + 2y = 17$;
$5x = 3y - 1$.

12. $x = 5y - 11$;
$y = 6x - 21$.

13. $3x - \dfrac{y}{2} = 11$;
$\dfrac{x}{2} + y = 10\tfrac{1}{2}$.

14. $8x - 6y = 5$;
$2x + 4y = 9\tfrac{1}{2}$.

15. $3x - 2y = 0$;
$5x - 3y = 2$.

16. $4x - 3y = 9$;
$\dfrac{x}{y} = 1\tfrac{1}{2}$.

17. $x - 1\tfrac{1}{2}y = \tfrac{1}{4}$;
$5x + 3y = 17$.

18. $2x - 3y = -8$;
$x + \tfrac{2}{3}y = 9$.

19. $5x - 3y = 1$;
$2x + 2\tfrac{1}{3}y = 11$.

20. $6x - 4y = 23$;
$\dfrac{x + 2\tfrac{1}{2}}{4} = y - \tfrac{1}{2}$.

119. Problems involving simultaneous equations.

Problems which involve finding the value of two unknown quantities may be solved by simultaneous equations.

EXAMPLE 1. Five coins of one kind and eight of another amount to £1. 8s. 6d., while eight coins of the first kind and five of the second kind amount to £1. 10s. What are the coins?

Let $\quad x =$ value of first kind of coin in sixpences

$y = \quad$,, $\quad$ second $\quad$,, $\quad$,, $\quad$,,

Then $\qquad 5x + 8y = £1.\ 8s.\ 6d. = 57$ sixpences,

and $\qquad 8x + 5y = £1.\ 10s.\ 0d. = 60 \qquad$,,

$$\therefore \quad \left.\begin{array}{l} 5x + 8y = 57 \\ 8x + 5y = 60 \end{array}\right\}.$$

Solving these equations we get $x = 5$, $y = 4$.

$\therefore \qquad\qquad$ first coins are halfcrowns,

and $\qquad\qquad$ second ,, florins.

EXAMPLE 2. If five first-class tickets and 24 third-class tickets cost £6. 4s. 2d., and the price of 3 first-class tickets decreased by the price of 5 third-class ones amounts to 9s. 10d., find the cost of each ticket.

Let $\quad x =$ the price in shillings of first class,

$y = \quad$,, $\quad$,, $\quad$ third ,,

Then $\qquad 5x + 24y = £6.\ 4s.\ 2d. = 124\tfrac{1}{6}s.$;

$\qquad 3x - 5y = 9s.\ 10d. = 9\tfrac{5}{6}s.$

Thus $\qquad\qquad 5x + 24y = 124\tfrac{1}{6}$,

and $\qquad\qquad 3x - 5y = 9\tfrac{5}{6}$.

Solving these equations we get $x = 8\tfrac{5}{6}$, $y = 3\tfrac{1}{3}$.

$\therefore \qquad\qquad$ first-class ticket costs 8s. 10d.,

and $\qquad\qquad$ third ,, ,, 3s. 4d.

Exercises 54

1. The sum of two numbers is 14, and four times the greater exceeds ve times the less by 2. Find them.

2. Divide 112 into two parts, proportional to 5 and 9.

3. One number is greater than another by 21. If the greater is ivided by six times the smaller, the quotient is 1, and remainder 6. 'ind the numbers.

4. Four chisels and three saws cost £2. 7s. 0d., and six chisels cost s. 4d. more than seven saws. What is the price of a chisel?

5. In the equation $E = aL + b$, we are given that when $E = 10$, $L = 25$, nd when $E = 16$, $L = 40 \cdot 81$. Find the values of a and b, and find the alue of E when $L = 40$.

6. If two first-class tickets between London and Edinburgh cost s. 6d. more than three third-class, and five thirds cost £2 more than hree firsts, find the price of each ticket.

7. 250 attend a concert, some pay 2s. 6d., and some 1s. 6d. per ticket. f the total receipts amount to £21. 10s. 0d., how many tickets of each ind were sold?

8. A man is paid 5s. 6d. for every day he works, and is fined sixpence or each day's absence. After 28 days, the amount due to him is £6. 4s. 0d. [ow many days was he absent?

9. The distance between two towns is 16 miles. A person travels artly by rail and partly by road. The fare by rail is $1\frac{1}{4}d.$ per mile, and y road $\frac{3}{4}d.$ a mile. The total fare is 1s. 6d. Find the distance travelled y each mode of conveyance.

10. Two quantities X and Y are connected by a law in the form of $= aX + b$. When $X = 3$, $Y = 9$. When $X = 1$, $Y = 5$. Find the connection etween a and b.

11. The size across the flats of a hexagonal nut for a 1 in. diameter olt is found to be $1\frac{5}{8}$ in. The corresponding size of a 2 in. bolt is $3\frac{1}{8}$ in. tate the law connecting these quantities if it is in the form of $F = aD + b$. tate also the size across the flats of a $3\frac{1}{2}$ in. nut.

12. If 5 men and 24 boys earn £61. 10s. in a week, and if the wages of boys per week is 7s. 6d. more than those of a man, find the weekly age of a boy and man respectively.

13. A sum of £5. 12s. 6d. is paid partly in sixpences, and partly i florins. If the total number of coins is 81, how many of each are there

14. A number consisting of two digits is 3 less than five times the sum of its digits. If 18 is added to the number the digits will be reversed Find the number.

15. Find the distance between two stations, when by increasing th speed 10 m.p.h. a train can do the journey in 2 hours less, and by decreasing the speed 6 m.p.h., it takes the train 2 hours longer.

CHAPTER XIII

CO-ORDINATES, STRAIGHT LINES AND SLOPES

120. It has already been shown in para. 116 how the relationship between two quantities which depend on one another for their values may be shown on a graph.

This graphic method will now be extended to the solution of equations.

121. Position of a point.

The position of a point in a plane may be exactly determined if its perpendicular distances from two straight lines at right angles are known. The two straight lines are called **Lines of Reference** or **Axes**.

EXAMPLE 1. Find the position of a point which is 1 in. away from and to the right of a vertical line, and 2 in. away from and above a horizontal line.

Draw the axes and call them OX and OY (Fig. 81). Draw the line AB 1 in. away from OY and to its right, and draw the line CD 2 in. away from OX and above it. The two lines intersect at P. Then this is the point fulfilling the required conditions.

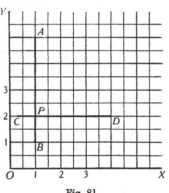

Fig. 81

122. Co-ordinates of a point.

The two measurements which determine the position of a point are called its **co-ordinates**. Co-ordinates are generally expressed in brackets. Thus the above co-ordinates in Example 1 would be expressed as (1, 2).

The first co-ordinate is always the number of units from the OY axis, and the second the number of units away from the OX axis. The intersection of the axes is always known as the **origin**.

123. Signs of direction.

So far measurements to the right of the vertical, and above the horizontal axes only have been considered, but direction to the left and downward may also be considered. To denote which direction is meant, conventional signs are used, namely direction to the right and upward are regarded as positive directions, and to the left and downward as negative directions.

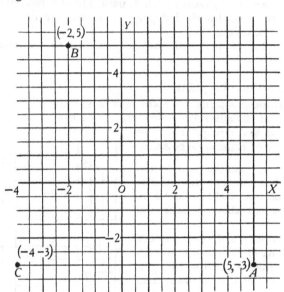

Fig. 82

EXAMPLE 2. Plot the following points:

$$A\ (5,\ -3),\ B\ (-2,\ 5),\ C\ (-4,\ -3)\ (\text{Fig. 82}).$$

124. Position of a line.

The position of a straight line may be determined if the co-ordinates of its extremities are known.

EXAMPLE 3. Draw the straight line, the co-ordinates of the extremities being $(2, 6), (3, -4)$ (see Fig. 83).

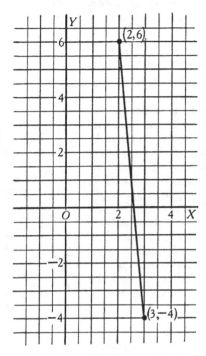

Fig. 83

Exercises 55

Plot the following points:

1. $(3, 3)$, $(-8, 2)$, $(0, 4)$, $(1, -4)$.

2. $(-2, 0)$, $(3, 6)$, $(-5, -2)$, $(5, -3)$.

Draw the following lines:

3. (a) $(2, 6)$, $(-3, -4)$; (b) $(4, -2)$, $(0, 5)$.

4. (a) $(-3, 2)$, $(5, 5)$; (b) $(-2, 0)$, $(0, 5)$.

5. Plot the four points $(2, 1)$, $(6, 6)$, $(7, 1)$, $(2, 9)$.

Draw the two diagonals, and find the co-ordinates of their intersection.

6. Construct the following quadrilateral: (3, -4), (-0.6, -2 (-3, 2), (4, 6). Find the co-ordinates of the points of intersection of th diagonals.

7. Construct the triangle A (3, 6), B (6, -1), C (-4, -5). Find th co-ordinates of mid-point of AB.

125. Mark the point (2, 4) on your paper and join it to the origin so obtaining OP (Fig. 84).

Now if any point is taken on this line OP, the perpendicular from this point to the OX axis will be found to be twice the length of the base intercepted between the foot of the perpendicular and the origin, viz. $PF = 2FO$; $p_1f_1 = 2f_1O$; $p_2f_2 = 2f_2O$; $p_3f_3 = 2f_3O$. All lines perpendicular to the OX axis are called *y lines* and those parallel to the OX axis are called *x lines*. The y lines are more usually called **ordinates** and the x lines **abscissae**.

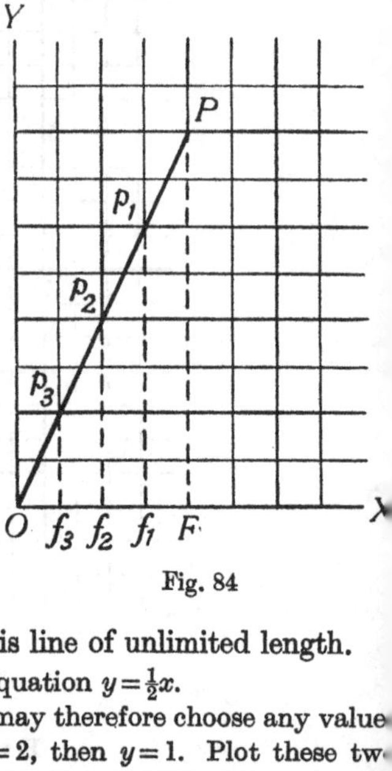

The relation existing between the ordinates and abscissae in Fig. 84 may be written
$$y = 2x.$$
Therefore the straight line OP (Fig. 84) may be represented by the equation

Fig. 84

$y = 2x$. This equation will represent this line of unlimited length.

EXAMPLE 1. Draw the graph of the equation $y = \frac{1}{2}x$.

As x and y are variable quantities we may therefore choose any value we wish. Put $x = 8$, then $y = 4$; put $x = 2$, then $y = 1$. Plot these two points (8, 4) and (2, 1). Join them. Then the line AB (Fig. 85) represent the equation $y = \frac{1}{2}x$.

EXAMPLE 2. On the same axes draw the graphs of the equation (i) $y = 2x + 3$, (ii) $y = x - 3$.

(i) $y = 2x + 3$.

x	0	1
y	3	5

(ii) $y = x - 3$.

x	3	5
y	0	2

Choose values for x in both equations, and draw both lines AB an CD (Fig. 86).

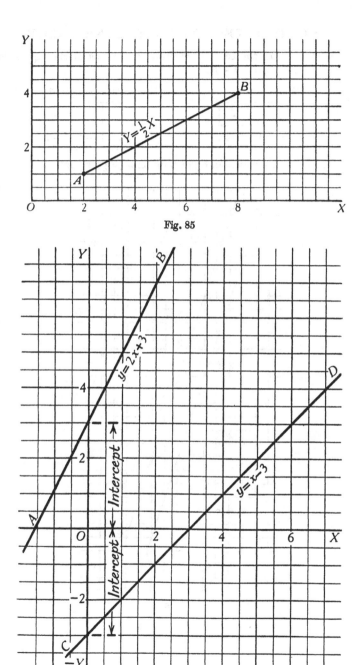

Fig. 85

Fig. 86

EXAMPLE 3. On the same axes draw the graphs of the equations (i) $y = 2 - 3x$; (ii) $y = -1 - 3x$.

(i) $y = 2 - 3x$.

x	1	2
y	-1	-4

(ii) $y = -1 - 3x$.

x	1	2
y	-4	-7

Choose values as shown and plot both lines (Fig. 87) EF, HK.

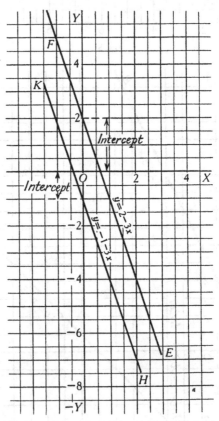

Fig. 87

Now compare the graphs in Figs. 86 and 87. It is seen that in Fig. 86 the **slopes** of the two lines are different, while in Fig. 87 the two lines are parallel (same slope). Looking at the equations, the *coefficients of x* in the first two are different, while in the second two equations they are alike (equal).

The coefficient of *x* determines the slope of the line. The larger the coefficient the steeper the slope.

126. Again it will be noticed that the two lines in Fig. 86 cut the *X* axis at an acute angle, while the lines in Fig. 87 cut the *OX* axis at an obtuse angle. In the equations in Fig. 86 the sign of the coefficient of *x* is positive, while in Fig. 87 it is negative. **If the sign of the coefficient of *x* is positive the line cuts the *OX* axis at an acute angle, if the sign is negative it cuts at an obtuse angle.**

It follows then, that if in two equations the signs are opposite, while the coefficients of *x* are equal, then the lines are perpendicular to one another. For example, $y = 2x + 4$ and $y = 4 - 2x$ represent two lines at right angles to each other.

127. In the equations $y = 2x + 3$, $y = x - 3$, $y = 2 - 3x$, $y = -1 - 3x$, each has a term independent of the value of *x*, viz. $+3$, -3, $+2$, -1. These are called *independent terms* or *constants*. It will be noticed from the graphs that **where the independent term is positive, the line cuts the *OY* axis above the origin, and where negative below the origin.**

The part of the *OY* axis between the origin and where the line cuts it, is called the **intercept**.

128. Generally the equation of a straight line may be one of the following:

$$\text{(i) } y = mx + c, \qquad \text{(ii) } y = -mx + c,$$
$$\text{(iii) } y = mx - c, \qquad \text{(iv) } y = -mx - c,$$

m being the coefficient of *x* and *c* the independent term, generally called the **constant** (Fig. 88).

Note. The value of *y* only must be expressed in terms of the other quantities. For instance if the equation were

$$3y = 6x + 4,$$

it would have to be changed to

$$y = 2x + \tfrac{4}{3}.$$

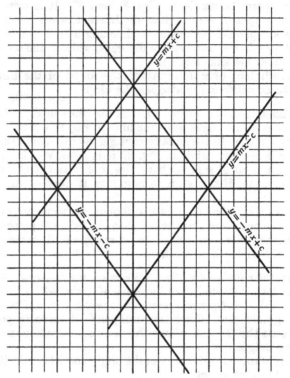

Fig. 88

Exercises 56

1. On the same axes plot the following equations, using either a half inch or a centimetre as your unit:

(i) $y = x$; (ii) $y = 2x$; (iii) $y = \frac{1}{2}x$.

2. Plot the following on the same axes, same scale as in previous question:

(i) $y = 2x + 3$;   (ii) $y = 2x - 3$;

(iii) $y = -2x + 3$;   (iv) $y = -2x - 3$.

3. Plot $2y = 3x + 4$, $y = 1\frac{1}{2}x - 3$.

4. Draw the graphs of $2y = x + 4$, and $y = 2x - 1$. What are the co-ordinates of the point of intersection of these two lines?

5. Draw the graphs of the following lines:

(i) $3x + 2y = 13$;   (ii) $4x = 3y + 6$;   (iii) $x - y = 1$.

f they have a common point of intersection, what are its co-ordinates

129. Graphical solution of an equation.

Consider the equations:

(1) $y = 2x - 5$;

(2) $y = x - 1$.

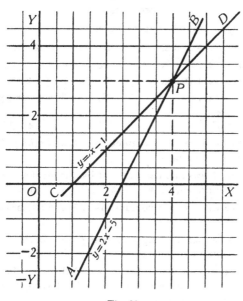

Fig. 89

Many pairs of values could be found for x and y which would atisfy Equation (1), but *only one pair* which would satisfy both quations *at the same time*.

Plot Equation (1) (Fig. 89). The co-ordinates of any point on the ne AB will satisfy the equation

$$y = 2x - 5.$$

Now draw the graph of equation

$$y = x - 1 \quad \text{(Fig. 89)}.$$

This gives the line CD. The two lines AB and CD cut at P. The point P lies on *both* lines, therefore its co-ordinates must satisfy both equations. The co-ordinates are $(4, 3)$, therefore $x = 4$ and $y = 3$ is the solution of the equation.

130. Equations which contain the first powers only of x and y are called **Linear Equations**, because the graph of any such equation is a **straight line**.

Exercises 57

Solve by graphs the following equations:

1. $5x - 3y = 1$; $x + 2y = 8$.
2. $3x + 4y = 18$; $3x - y = 3$.
3. $4x - 3y = 6$; $x + 2y = 7$.
4. $2x + 3y = 7$; $y - 2x = 5$.
5. $4x + y = 7$; $3x + 4y = 2$.
6. $3x - 2y = -2$; $5x + 2y = 18$.
7. $4x - 6y = 0$; $x + y = -5$.
8. $4x - 3y = 1$; $2x + y = 8$.
9. $\frac{x}{2} + y = 1$; $\frac{x}{y} = -3$.
10. $\frac{2x}{3} - y = 0$; $\frac{x}{4} + \frac{y}{2} = 7$.

11. One-third the sum of two numbers is 4, and one-half their difference is 2. Find the numbers.

12. The sum of two numbers divided by the smaller equals $2\frac{1}{4}$, and three times their difference is 6. What are the numbers?

13. One-third the sum of two numbers is 5, and twice their difference is 6. Find them.

14. Divide a bar 21 ft. long into two parts, so that $1\frac{1}{2}$ times the larger part will equal twice the smaller.

15. Twice the smaller of two numbers added to one-third the larger is 12, and four times the larger increased by 1 equals five times the smaller. What are the numbers?

131. To find the equation of a straight line.

In para. 128 it was seen that in general the equation of a straight line was of the form

$$y = mx + c.$$

If then the position of a straight line is known, two points can be taken on it and the co-ordinates of these points substituted in the general equation.

Thus the values of m and c may be found, and thence the actual equation for the line.

EXAMPLE 1. Find the equation of the straight line passing through the points $(1, -3)$, $(4, 3)$.

Let $y = mx + c$ be the equation. Since it passes through the point $(1, -3)$ we can substitute these two values for x and y in the equation.

$$\therefore \quad -3 = m + c.$$

Likewise we can substitute the co-ordinates $(4, 3)$ in the general equation.

$$\therefore \quad 3 = 4m + c.$$

Therefore by solving the simultaneous equations:

$$-3 = m + c,$$
$$3 = 4m + c,$$

we get $m = 2$ and $c = -5$.

Therefore the equation of the line passing through the given points $(1, -3)$, $(4, 3)$ is

$$y = 2x - 5.$$

EXAMPLE 2. Find (a) the slope, (b) the intercept, (c) the equation of the straight line passing through $(2, 8)$, $(4, 14)$.

$$y = mx + c.$$
$$\therefore \quad 8 = 2m + c.$$
$$\therefore \quad 14 = 4m + c,$$

hence $\qquad 3 = m$, and $2 = c$.

Therefore the equation of the line is

$$y = 3x + 2,$$

the slope of the line $= 3$,

the intercept on the OY axis is $+2$.

(*For notes on slope and intercept revise paras. 125 and 127.*)

Exercises 58

1. Find the equations of the straight lines through the following pairs of points:

(i) $(3, 2)$ and $(-2, 3)$;   (ii) $(-1, 2)$ and $(3, 0)$;

(iii) $(-4, 0)$ and $(2, -3)$;   (iv) $(1, 4)$ and $(-1, 0)$;

(v) $(4, -3)$ and $(-2, 6)$;   (vi) $(1, 0\cdot5)$ and $(4, -4)$;

(vii) $(3, 1\cdot2)$ and $(5, 0\cdot4)$;   (viii) $(2, 1\frac{1}{2})$ and $(5, 7\frac{1}{2})$.

144 CO-ORDINATES, STRAIGHT LINES, SLOPES

2. Plot the points given below. Draw the line which lies most evenl
among them and find its equation.

x	2	6	8	11	15	17	18
y	1	3	4	6	8	9	10

3. For the differential pulley the following are corresponding value
of the weight (W), and the effort (P) required to raise it:

W lb.	80	120	200	240	300	350	400	450
P lb.	13	18	28	33	40·5	47	53	59

Plot these and obtain the law of the form $P = a + bW$, connecting
and W.

4. The following readings were plotted. Find the equation connectin
x and y.

x	1	2	3	4	5	6
y	0·5	2	3·5	5	6·5	8

5. The table shows the relationship between weights (W) in lb
raised by a screwjack, and the required force (P) in lb. required:

W lb.	0	20	40	60	90
P lb.	0·9	1·9	2·9	3·8	5·3

Find the law connecting P and W, and find what effort will be require
to raise 80 lb.

6. A spring was stretched by addition of weights. The following ar
the successive weights and lengths:

Load in lb. (P)	3	6	10	15	21
Length in ft. (L)	15	21	29	39	51

Find the law of the form $L = a + bP$, connecting load and length.

7. The following series of readings connect x and y. Draw a graph, nd find the equation connecting x and y.

x	4	6	8	10	12
y	3	6	9	12	15

8. Plot the following values of x and y, and find the relation of the rm $y = mx + c$:

x	1	2	3	4	5	6	7
y	5·4	4·75	4·15	3·55	2·95	2·35	1·8

9. Graph to show the relationship between the *circumference and iameter of a circle*. Complete the following table:

D	1·0	1·5	2·0	2·5	2·75	3·0
C	3·14					

Plot circumference vertically and diameter horizontally and draw he straight line graph. Find its slope. Write down the law of the graph.

10. The size across the flats, or the "spanner size" of a Whitworth ut is shown approximately in the following table:

Diameter	0·5	0·75	1·0	2·0	3·0
Size across flats	0·875	1·25	1·625	3·125	4·625

Plot a graph *Flats* v. *Diameters*. Find the law. It is in the form of $^\prime = aD + b$.

11. The readings on a Centigrade thermometer were compared with hose on a Fahrenheit thermometer and the following readings found:

C.	10°	25°	35°	50°	70°	80°	100°
F.	50°	75°	98°	121°	160°	174°	212°

Find the law connecting Fahrenheit and Centigrade readings, i.e. ot the graph F. v. C.

12. The following table gives D (the outside diameter of a Whit worth bolt) and d the core diameter:

D	0·25	0·5	0·75	1·0	1·5
d	0·186	0·393	0·622	0·84	1·287

Find the law connecting core diameter and bolt diameter.

CHAPTER XIV

MENSURATION OF SOLIDS

132. We saw in para. 13 that the amount of space occupied by a ·lid is called its volume.

A **prism** is a solid the ends of which are parallel to each other and ·ave the same shape. The *axis*, or centre-line, of a prism may be ·ertical to its end faces. In this case we have a *right prism*. If the ·is is inclined at an angle other than 90° we have an *oblique prism*.

A **pyramid** is a solid having for its base any rectilineal plane ·gure and for its sides triangles which meet at a point known as the ·rtex or *apex* of the pyramid. In the case of a *right pyramid* the ·rtex lies immediately above the centre of the base and the axis is ·rtical. In an *oblique pyramid* the axis is inclined to the plane con-·ining the base.

Both prisms and pyramids are named according to the shape of ·e base. Thus we have triangular, square, pentagonal, etc., prisms ·d pyramids. In Fig. 90 we have right and oblique square prisms ·d pyramids.

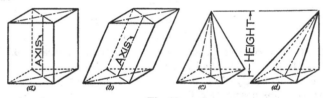

Fig. 90

133. A **cube** is a right prism having length, breadth and thickness ·ual. A **cubic inch** is the volume of a cube of which each edge is ·in. long. A **cubic foot** is the volume of a cube of which each edge is ·1 ft. long. The cubic inch and cubic foot are common **units of ·lume**. *The table of cubic measure* is deduced from that of linear ·easure and Fig. 91 should make this clear. Let *ABCD* be a square ·ch side of which is 1 yd. or 3 ft. long. Then its area is 9 sq. ft. If ·om *ABCD* perpendiculars be drawn each 3 ft. long, we can readily ·mplete the cube shown at (*b*). If each face of the cube be divided into 9 equal parts we shall have three layers of cubes—each layer consisting of 9 small cubes. The total number of small cubes is thus 27, and each is of unit volume, viz. 1 cu. ft. (see (c)).

Vol. of cube of 1 ft. edge $= 1^3 = 1$ cu. ft.

,, ,, 3 ft. ,, $= 3^3 = 27$ cu. ft. $= 1$ cu. yd.

,, ,, s ft. ,, $= s^3$ cu. ft. $= \dfrac{s^3}{27}$ cu. yd.

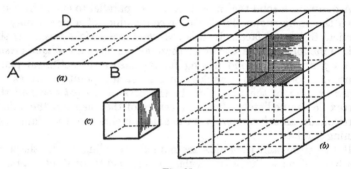

Fig. 91

Useful table

1728, i.e. 12^3, cu. in. = 1 cu. ft.

27, i.e. 3^3, cu. ft. = 1 cu. yd.

1 cu. ft. of water weighs approximately 1000 oz. (more accurately $62\frac{1}{3}$ lb.)

1 gall. holds 10 lb. of distilled water.

1 gall. contains nearly $277\frac{1}{2}$ cu. in.

1 cu. ft. contains nearly $6\frac{1}{4}$ gall.

134. The general case of the right prism.

1. Consider the rectangular prism shown in Fig. 92 and suppose a slice or slab of unit thickness cut off parallel to *ABCD*.

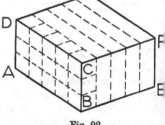

Fig. 92

2. For every whole *unit of area* in this face there will be a *unit of volume* in the slab, and for every fractional unit of area a fractional unit of volume.

Suppose the edge AB is 5 ft. and the edge BC 3 ft. Then the area
of the face $ABCD$ will be 15 sq. ft. and the volume of the first slab
will be 15 cu. ft.

As there are 5 slices ($BE=5$ ft.) the volume of the whole prism is
$\times5\times5=75$ cu. ft.

3. Thus the **volume of a right prism**

$$= \text{length} \times \text{breadth} \times \text{thickness},$$

r, generally,   **= area of end × length.**

If the same unit is used for length, breadth, and thickness the volume
will be in corresponding cubic units.

135. Volume of a cube.

We have seen that the length, breadth and height of a cube are
qual. Since all its edges are equal they may each be represented by l.
Then   Volume $(V) = $ area of end $\times$ vertical height

$$= l \times l \times l = l^3.$$

Thus   **Volume $=$ (edge)^3; edge $= \sqrt[3]{\text{volume}}$.**

EXAMPLES. (1) Find the volume in cubic inches of each of the prisms
shown in Fig. 93.

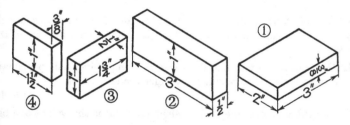

Fig. 93

Vol. of prism $1 = 2 \ \times 3 \ \times \frac{3}{8} = 2\frac{1}{4}$ cu. in.

,, ,, $2 = 3 \times \frac{1}{2} \times 1 = 1\frac{1}{2}$,,

,, ,, $3 = \frac{1}{2} \times 1\frac{3}{4} \times 1 = \frac{7}{8}$,,

,, ,, $4 = 1\frac{1}{2} \times \frac{3}{8} \times 1 = \frac{9}{16}$,,

(2) A rectangular tank is 5 ft. long, 2 ft. 6 in. wide, holds 100 gall.
nd the depth of water in the tank. (1 gall. of water weighs 10 lb.;
cu. ft. of water weighs 62·3 lb.)   (*U.E.I.*)

1 gall. of water weighs 10 lb., therefore 100 gall. weigh 1000 lb.

Then $\qquad$ no. of cu. ft. in tank $= \dfrac{1000}{62 \cdot 3}$.

Let $d =$ depth of water (in ft.).

Then $\qquad$ volume of water $= 5 \times 2\frac{1}{2} \times d$ (cu. ft.).

$$\therefore \quad 5 \times 2\tfrac{1}{2} \times d = \frac{1000}{62 \cdot 3}.$$

$$\therefore \qquad d = \frac{1000}{62 \cdot 3} \div 12\tfrac{1}{2} = 1 \cdot 28 \text{ ft.}$$

Exercises 59

1. The edge of a cube is 12 in. long. State its volume (*a*) in cubic inches, (*b*) in cubic feet.

2. A rectangular solid has a base of 4 ft. by 3 ft. and is 5 ft. 6 in. high. What is its volume?

3. The area of the base of a hexagonal prism is 6·5 sq. cm. Its height is 8·5 cm. What is its volume?

4. The floor of a room is 80 ft. by 40 ft. It may be of oak $1\frac{1}{8}$ in. thick, or of deal $1\frac{1}{2}$ in. thick. If deal costs 3*s.* and oak 6*s.* per cubic foot, find the cost in each case.

5. A cubic foot of elm weighs 40·32 lb., and 160 planks, each 8 ft. long by 9 in. broad, together weigh 1·35 tons. Find the thickness of the planks. *(C. and G.)*

6. Given that sheet lead, of thickness 0·135 in., weighs 8 lb. per square foot, find to the nearest pound the weight of a cubic foot of lead.

(R.S.A.)

7. If a cubic foot of steel weighs 489·6 lb., what is the weight of a bar 1 in. square by 1 ft. long?

8. A roof measures 48 ft. by 20 ft. All the rainwater falling upon it is collected into an empty rectangular tank 4 ft. by 2 ft. 6 in. What will be the depth of water in the tank after a rainfall of $\frac{1}{10}$ in.? (A rainfall of $\frac{1}{10}$ in. may be supposed to cover the roof to a uniform depth of $\frac{1}{10}$ in.)

136. Surface area of a rectangular solid is found by adding the areas of the six rectangles comprising its surface.

Surface area of any prism is found by taking the sum of the areas of the ends and the areas of the lateral surfaces.

Fig. 94 shows the development of a rectangular prism. It has three pairs of parallel surfaces, each of which is rectangular.

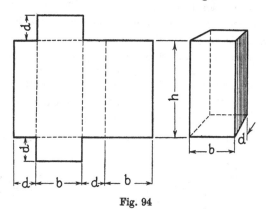

Fig. 94

Let height of prism $= h$ units

  ,, breadth ,, $= b$,,

  ,, depth   ,, $= d$,,

(1)   Total surface area $= 2hb + 2hd + 2bd$

$$= 2\,(hb + hd + bd) \text{ square units.}$$

(2) Or, perimeter of base $= 2\,(b + d)$,

    $\therefore$ area of lateral faces $= 2\,(b + d) \times h$,

and   area of end faces $= 2\,(bd)$,

    $\therefore$ whole surface $= 2\,(b + d)\,h + 2bd$ square units.

Exercises 60

1. Find the surface area of a cube having an edge of 15 in.

2. A room is 18 ft. long by 15 ft. wide by 10 ft. high.

    (a) Find the area of the four walls in square feet.

    (b) Assuming that 10 p.c. of this area is occupied by cupboards, windows, etc., find the cost of distempering the remainder at 6d. per square yard.   (R.A.F.)

3. Find the cost to the nearest shilling of lining a cistern 10 ft. 3 in. long by 6 ft. 6 in. wide by 5 ft. 4½ in. deep with lead at 21s. per cwt. which weighs 8 lb. per square foot. Include the lid.   (C. and G.)

4. An open-top water tank is 3 ft. long by 1 ft. wide by 18 in. deep. It is made from 16-gauge mild steel plate. Find the weight of sheet metal required to make the tank, assuming that 20 p.c. extra is required to make laps, seams, etc. 1 sq. ft. of this plate weighs $2\frac{1}{2}$ lb.

5. A roof consists of:

26 lengths of 4 in. by 2 in. common rafters, 18 ft. long
2   ,,   9 in. by 4 in. purlins,   36 ft. ,,
1   ,,   9 in. by 2 in. ridge,   36 ft. ,,

Find the cost of the timber if it costs £28 per standard.

(A standard = 165 cu. ft.)

6. Find the volume of the slotted casting shown in Fig. 95.

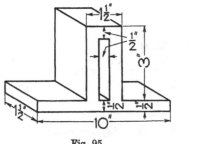

Fig. 95

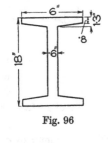

Fig. 96

137. Miscellaneous prisms.
Volume = area of end × length.

We assume that a prism has the same cross-section throughout its length and that its ends are parallel.

Exercises 61

1. A swimming bath is 80 ft. long by 20 ft. wide by 9 ft. deep at one end and 4 ft. deep at the other. How many cubic feet of water does it contain?

(*This may be considered a trapezium-shaped prism which is 20 ft. long.*)

2. A swimming bath is x yd. long and y yd. broad; the water is a ft. deep at one end and b ft. deep at the other; the floor of the bath slopes uniformly. Find a formula for the number of cubic feet of water in the bath.

What is the volume of water in such a bath which is 30 yd. long,
2 yd. broad, when the water is 3 ft. 6 in. deep at one end and 6 ft. 6 in.
deep at the other?   (*N.C.T.E.C.*)

3. Fig. 96 shows a cross-section of a joist. What is its weight per foot
un? (Take 1 cu. in. of steel to weigh 0·28 lb.)

4. A trench in section is a symmetrical trapezium. At the bottom
is 8·6 ft. wide, at the top 12·5 ft. wide, and it is 3·5 ft. deep. If it is
) yd. long, how many cubic yards of earth were excavated in digging
? Neglect waste, etc.

138. Right circular cylinder.

Suppose *ABCD* (Fig. 97) is a rectangular piece of cardboard having
knitting needle fastened by adhesive tape along the edge *AD*. If
e knitting needle be held vertically and the rectangle *ABCD* rotated
out it as an axis, *the rectangle will sweep out a solid figure of re-
lution called a right circular cylinder.*

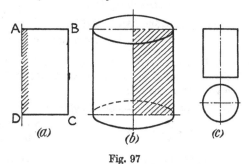

Fig. 97

The ends of the cylinder are plane figures, e.g. circles. The axis joins
eir centres and, of course, is vertical. We may consider this
linder as a circular prism.

An **oblique cylinder** is one in which the axis is not perpendicular
the ends. Whenever in this book the word *cylinder* is used a right
linder can be assumed.

139. Volume of a cylinder.

We may consider a cylinder as a right prism having an infinite
imber of sides. Thus

$$\text{volume} = \text{area of end} \times \text{length.}$$

Let h = vertical height; d = diameter of base; r = radius of base. Then

$$\text{area of end} = \frac{\pi}{4}\, d^2, \text{ or } \pi r^2,$$

$$\therefore \quad \textbf{Volume of cylinder} = \frac{\pi}{4}\, d^2 h, \text{ or } \pi r^2 h.$$

WORKED EXAMPLES.

1. A circular table has a top 7 ft. in diameter and 1 in. thick. (a) What is the volume of the top? (b) How much would it cost to polish the top surface at 9d. per square foot?

(a) Use the formula
$$\begin{aligned} V &= 0{\cdot}7854 d^2 h \\ &= 0{\cdot}7854 \times 49 \times \tfrac{1}{12} \\ &= 3{\cdot}21 \text{ cu. ft. (approx.).} \end{aligned}$$

(b) Area of top
$$\begin{aligned} &= 0{\cdot}7854 \times 49 \\ &= 38{\cdot}485 \text{ sq. ft.} \end{aligned}$$

Cost at 9d. per sq. ft.
$$\begin{aligned} &= 38{\cdot}485 \times 9 = 346{\cdot}4d. \\ &= \text{£1. 8s. 10d. (to nearest penny).} \end{aligned}$$

2. A cylinder of 8 in. diameter just fits into a cubical box of 8 in. edge. (a) Find the number of cubic inches void. (b) Write this as a percentage of the volume of the cube.

(a)   Void = vol. of cube − vol. of cylinder

$$= 8^3 - \frac{\pi}{4} \times 8^2 \times 8$$

Note that 8^3 is a common factor

$$= 8^3 \left(1 - \frac{\pi}{4}\right)$$

$\dfrac{\pi}{4} = 0{\cdot}7854$

$$= 8^3 (0{\cdot}2146)$$
$$= 512 \times 0{\cdot}2146$$
$$= 109{\cdot}875 \text{ cu. in.}$$

(b)   Percentage void $= \dfrac{\text{Vol. of void}}{\text{Vol. of cube}} \times 100$

$$= \frac{109{\cdot}875 \times 100}{512}$$
$$= 21{\cdot}46 \text{ p.c.}$$

Exercises 62

$$\left(\frac{\pi}{4} = 0.7854\right)$$

1. A circular sheet of lead is 18 in. in diameter. What is the area of a rcular face? It is $\frac{1}{4}$ in. thick. What is its weight if lead weighs 0·4 lb. er cubic inch?

2. Ten rollers are required. The diameter of each is to be $2\frac{1}{2}$ in., and e length of each 5 in. Find the total volume in cubic inches.

3. The largest possible cylinder is made from a cube, the edge of hich is 7 in. What fraction of the wood is wasted?

4. A cylinder is 21 ft. in diameter. It collects rain from a roof whose ea is 1172 sq. ft. What will be the depth of water in the tank after in. of rain? ($\pi = 3\frac{1}{7}$.)

5. A cylindrical pillar is 12 in. in diameter and 38 ft. high. If its eight is 350 lb., find the weight of 1 cu. in. of the material.

6. A piece of plate is 12 in. by 6 in. by 1 in. and eight holes of $\frac{7}{8}$ in. ameter are drilled through its flat face. What is the resulting loss in olume? What is the volume of the drilled plate?

7. A piece of metal is hexagonal in shape and 1 in. thick. A hole 1 in. diameter is drilled through its centre as in a nut. If the distance across e corners of the hexagon is 2 in., find the volume of metal remaining.

(U.E.I.)

8. A cylinder has a diameter of $3\frac{1}{2}$ in. and is partly filled with water. piece of lead 3 in. $\times 2\frac{1}{3}$ in. $\times 10\frac{1}{2}$ in. is placed in it so as to be com- etely immersed. How much higher in the cylinder will the water and? ($\pi = 3\frac{1}{7}$.)

9. A jig plate is triangular in plan, its sides being 6 in., 8 in., and in. Its thickness is $1\frac{1}{2}$ in. Two holes are drilled through the triangular ce their diameters being $\frac{3}{4}$ in. and $1\frac{5}{8}$ in. respectively. What is the eight of the plate if the metal weighs 0·26 lb. per cubic inch?

10. A cubic inch of copper is drawn out into wire of 0·04 in. diameter. nd to the nearest inch the length of the wire pposing all the copper could be used.

11. A cylindrical bar (Fig. 98) is 10 in. in ameter and is subjected to a tensile load of 0 tons. What is the tensile stress at the rcular section *CD*? Take stress as load ÷ area section.

Fig. 98

12. A petrol engine piston (Fig. 99 (*a*)) is $3\frac{1}{2}$ in. in diameter. What is the total downward force upon it at 350 lb. per square inch?

13. A $1\frac{1}{2}$ in. diameter eye-bolt is 1·29 in. in diameter at the core (or bottom of thread). Find the core area and, allowing a stress of 10,000 lb. per square inch, find the load it will carry (Fig. 99).

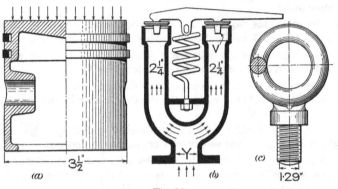

Fig. 99

14. A locomotive safety valve has two valves, each of $2\frac{1}{4}$ in. diameter held down by a central spring as shown in Fig. 99.

(*a*) Find to the nearest lb. the total stretching force on the spring when the steam is just blowing off at 120 lb. per square inch.

(*b*) Find the diameter Y, to the nearest round dimension, so that the cross-sectional area at Y equals the sum of the cross-sections of the valve seats.

140. Volume of a hollow cylinder (Fig. 100).

The volume of a hollow cylinder such as a pipe or washer consists of the difference between the volumes of cylinders of different diameters but having the same vertical height.

Let D, d be the internal and external diameters respectively, and L the length.

Volume = area of end × length.

Now the end of a hollow cylinder is an

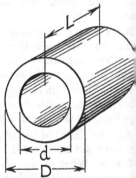

Fig. 100

annulus, and we saw in para. 114 that the area of an annulus could
be written

$$\text{area} = \pi \ (R+r) \ (R-r), \text{ or } \frac{\pi}{4} \ (D+d) \ (D-d).$$

Thus

Volume of hollow cylinder = π $(R+r)$ $(R-r)$ × L

$$= \frac{\pi}{4} \ (D+d) \ (D-d) \times L.$$

EXAMPLE. A metal pipe of circular section has an outer diameter of
$3\frac{3}{4}$ in., and the thickness of metal is $\frac{1}{8}$ in. If 1 cu. ft. of the metal weighs
170 lb., find the weight of 1 yd. of the pipe to the nearest ounce.

(*R.S.A.*)

Sectional area

$$= \frac{\pi}{4} \ (3\tfrac{3}{4} + 3\tfrac{1}{2}) \ (3\tfrac{3}{4} - 3\tfrac{1}{2})$$

$$= \frac{\pi}{4} \times \frac{29}{4} \times \frac{1}{4} = \frac{29\pi}{64} \text{ sq. in.}$$

Vol. of pipe 1 yd. long $= \dfrac{29\pi}{64} \times 36$ cu. in.

$$= \frac{29\pi \times 36}{64 \times 1728} \text{ cu. ft.}$$

∴   Weight

$$= \frac{29\pi \times 36 \times 170}{64 \times 1728} \text{ lb.}$$

$$= 5 \cdot 04 \text{ lb., say 5 lb. 1 oz. } Ans.$$

Exercises 63

$$\left(\frac{\pi}{4} = 0 \cdot 7854. \ \ Use \ logarithms \right)$$

1. Find the volumes of the following hollow cylinders:

(*a*) Inside diameter   = 15·8 in.
    Outside diameter = 22·2 in.
    Length   = 5 in.

(*b*) Inside diameter   = 25 in.
    Outside diameter = 50 in.
    Length   = 2 in.

2. A circular piece of zinc of thickness t in. and diameter D in. is
pierced by a hole of diameter d in. Write a formula for the volume of
the zinc in cubic inches.

3. Find the volume in cubic inches of the metal comprising a cast-iron pipe, the inside diameter of which is 3 in., outside diameter is 4 in., and length 6 ft. If 1 cu. in. of cast iron weighs 0·26 lb., find the weight of the pipe.

4. An iron washer is a flat perforated cylinder of $\frac{1}{8}$ in. thickness. The external diameter is $1\frac{1}{8}$ in., internal diameter is $\frac{1}{2}$ in. Given that 1 cu. ft. of the material weighs 483 lb., find, to the nearest ounce, the weight of a gross of these washers. (*R.S.A.*)

5. Fig. 101 shows the dimensions of a solid shaft and a hollow one, both made of steel weighing 485 lb. per cu. ft. Find the difference in weight between 6 ft. lengths of each shaft.

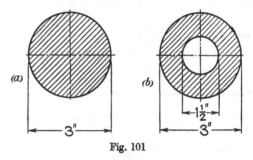

Fig. 101

6. Find the weight of the cast-iron flanged pipe shown in Fig. 102, if cast-iron weighs 0·26 lb. per cubic inch.

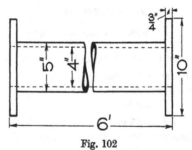

Fig. 102

141. The surface of a cylinder.

Take a short pasteboard tube or other simple cylinder and so cut it along a line parallel to its axis that its *curved surface can be spread ut flat.* The flat surface which results is rectangular in shape and is *ferred* to as *the development of the curved surface*. Its length equals *he* length of the cylinder whilst its breadth equals the circumference *of* the cylinder.

Area of curved surface

= (circumference of cylinder) × (length of cylinder)

$= \pi dl$, or $2\pi rl$.

Area of total surface

= area of curved surface + area of ends

$$= \pi dl + 2 \left(\frac{\pi}{4} d^2 \right)$$

$$= \pi d \left(l + \frac{d}{2} \right) = 2\pi r (r + l).$$

EXAMPLE. A cylindrical pipe is 15 in. diameter (inside), and 4 ft. *ng*. It is made from mild steel plate $\frac{1}{4}$ in. thick butt welded. Calculate *e* area of the plate required and also its weight if 1 sq. ft. of this plate *eighs* 10 lb.

The respective diameters are 15 in. and $15\frac{1}{2}$ in. Work to the mean *ameter*, i.e. $15\frac{1}{4}$ in. In a workshop it is customary to work in feet and *uare* feet on jobs of this size.

$$\text{Surface area} = \pi dl$$

$$= \frac{22}{7} \times \frac{15\frac{1}{4}}{12} \times 4$$

$$= 16 \text{ sq. ft. (nearly).}$$

Weight $\quad = 160$ lb.

Exercises 64

1. The commutator of a dynamo is a cylinder of 24 in. diameter and *... in.* length. Find the area of its curved surface in square feet and *uare* inches. $(\pi = 3\frac{1}{7}.)$

2. A cylindrical tank is closed at both ends and made of thin sheet *etal*. Base diameter is 3 ft. 6 in., and vertical height is 5 ft. 3 in. Find *e* surface area in square feet. $(\pi = 3\frac{1}{7}.)$

3. A condenser has 160 tubes each 10 ft. long and $\frac{3}{4}$ in. external *ameter*. Find the external cooling surface in square feet. $(\pi = 3\frac{1}{7}.)$

$(N.C.)$

4. The length of the core of a dynamo armature is 25 in., and it must have a radiating surface of at least 6000 sq. in. Determine its minimum diameter. Note that the core is cylindrical, the radiating surface being the cylindrical surface. ($\pi = 3\frac{1}{7}$.)   (*N.C.*)

5. The line diagram, Fig. 103, shows a sector portion of a circular disc. Draw the plan and elevation and determine its volume and surface area.

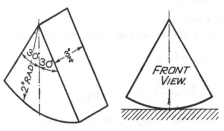

Fig. 103

142. Pyramids are solids bounded by plane faces. Whilst the base may be any rectilineal figure, the lateral faces are triangles—the vertices of which fall on one point called the *apex* of the pyramid. A *right pyramid* is one having its apex immediately above the centre of its base, i.e. its axis is perpendicular to the base. When the axis is not perpendicular to the base we have an *oblique pyramid*.

EXPERIMENTAL EXERCISE. Fig. 104 (*b*) shows a cube on the square base *EDCB*. Within the cube is shown an *oblique square pyramid* on the same base. Let us develop the surface of the pyramid. See Fig. 104 (*c*). Note that:

(*a*) The base *EDCB* is a square.

(*b*) The triangles *ABC* and *ABE* are right angled and isosceles.

By the theorem of Pythagoras $AC = AE = \sqrt{2} \times 4$.

(*c*) Obviously $AC = CA = AE$. Thus we draw the triangles *ACD* and *AED*, making $AC = AE = \sqrt{2} \times 4$.

The development should be cut out and folded up. Stiff drawing paper gives a good result if laps are allowed before cutting out.

If three such pyramids be folded up in this way it will be found that they can be arranged to form a cube.

Conclusion. The volume of the oblique pyramid $=\frac{1}{3}$ of the volume of
the cube.

If we imagined the oblique pyramid to be made up of an infinite
number of very thin sheets or planes (parallel to its base) we could, by
allowing them to slide over one another, change the oblique pyramid

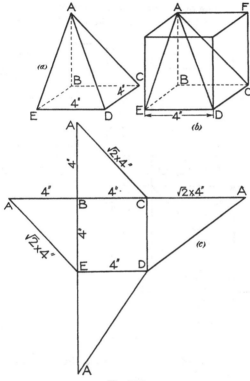

Fig. 104

to another standing on the same base but having the projection of its
apex within or without the base. However we slide the planes the
volume remains unaltered, hence for a right pyramid, as well as for an
oblique pyramid,

Volume $=\frac{1}{3}$ that of a prism of same height
     $=\frac{1}{3}$ **area of base × altitude.**

162   MENSURATION OF SOLIDS

Exercises 65

1. Find to three significant figures the weight of a leaden square pyramid if its base edges are each 4 in. long and its height is 8 in. (1 cu. in. of lead weighs 0·41 lb.)

2. Find the volume of a triangular pyramid the base of which is an equilateral triangle of 4 ft. edge and the vertical height is 12 ft.

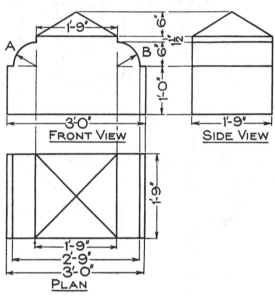

Fig. 105

3. Find the volume of a hexagonal pyramid each edge of the base being 2 in. and vertical height 8 in.

4. Calculate the weight of the block of Portland stone shown in Fig. 105, if the weight of 1 cu. ft. is 142 lb. The curves A and B are quadrants. (*N.C.*)

143. Surface areas of pyramids.

In Fig. 106, OA is the vertical height, AE is the length of the edge, AD is the perpendicular height of any one of the triangular faces.

Then the area of the *lateral surface* of this pyramid

$$= \text{area of } \triangle ACE \times 4$$
$$= \tfrac{1}{2}\,(3 \times 4) \times 4 = 24 \text{ sq. in.}$$

Area of whole surface $=$ lateral surface $+$ base

$$= 24 + 9$$
$$= 33 \text{ sq. in.}$$

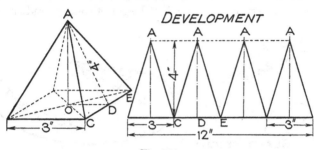

DEVELOPMENT

Fig. 106

144. The right cone.

The right cone is the solid figure of revolution generated or swept out by the rotation of a right-angled triangle about one of the sides containing the right angle.

The **base** of the cone is a circle.

Just as we regarded the cylinder as a prism the base of which has innumerable sides, so we may regard the cone as a pyramid, the base of which has an infinite number of very small sides. Thus

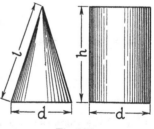

Fig. 107

Volume of cone
$$= \tfrac{1}{3} \,(\text{area of base} \times \text{altitude}).$$

Let $h =$ vertical height; $l =$ slant height; $d =$ diameter of base.

$$\therefore \quad \textbf{Volume} = \tfrac{1}{3}\,(0 \cdot 7854\ d^2 h), \quad \text{or} \quad \tfrac{1}{3}\,(\pi r^2 h).$$

In Fig. 107 we show a cone and a cylinder of the same vertical height and standing on equal bases.

Then   Volume of cone : volume of cylinder

as $\frac{1}{3}(\pi r^2 h) : \pi r^2 h$

or   1 : 3.

Thus the volume of the cylinder is 3 times that of the cone.

145. Surface area of a cone.

Experimental. On a piece of drawing paper draw a sector of a circle. (See *BOA* Fig. 108.) Cut out the sector and bend it round to form a cone. Observe that if $l=$ slant height of cone, $r=$ radius of its base,

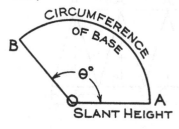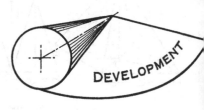

Fig. 108

Area of lateral surface = area of sector

= $\frac{1}{2}$ radius × arc of sector (para. 113)

= $\frac{1}{2}l \times 2\pi r$

= $\pi r l$.

Total surface area   = lateral surface + area of base

= $\pi r l + \pi r^2$

= $\pi r (l + r)$.

Exercises 66

1. Find the volumes of the following cones:

(a) Base = $1\frac{1}{2}$ in. diameter, vertical height = $1\frac{1}{4}$ in.

(b)   ,,   = $15\frac{1}{2}$ in.   ,,   ,,   = 12·6 in.

(c)   ,,   = 4 in.   ,,   ,,   = 5 in.

2. The base of a right cone is a circle of 3·25 in. diameter. Its vertical height is 5·24 in. If it is made of cast iron weighing 0·26 lb. per cubic inch, find its weight.

3. The base of a right circular cone is 4 in. diameter, and its height is in. Find the slant height and thence the area of the curved surface, *t* including the base.

4. A conical spire is 32 ft. diameter at the base and the slant height 108 ft. Find the cost of pointing it at an average cost of 2*s*. 3*d*. per juare yard. $(\pi = \frac{22}{7}.)$

5. Find to the nearest cubic foot the capacity of a bell tent con- sting of a cone upon a cylinder. Height of cone = 10 ft.; height of ·linder = 2 ft.; diameter of base = 16 ft.

6. A heap of coal takes the shape of a cone, the diameter at the bottom ·ing 108 ft. and slant height 90 ft.

If 42 cu. ft. of this coal weigh 1 ton, find the weight of coal in the heap. Give answer to the nearest ton. (*N.C.*)

7. The area of the curved surface of a right cone is 650 sq. in., and its ant height is 25 in. Find the diameter of its base. $(\pi = 3\frac{1}{7}.)$

8. Find the volume of the petrol tank shown in Fig. 109. It consists a cylinder and two cones.

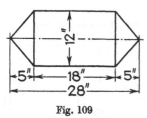

Fig. 109

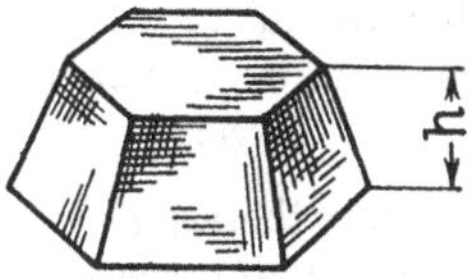

Fig. 110

146. Frustum of a pyramid.

If we cut across a pyramid along a plane parallel to its base we shall ave a *frustum* as shown in Fig. 110.

Volume of a frustum.

By a method too difficult to show at this stage it can be proved that e volume of a pyramid having parallel upper and lower faces can · written,

$$V = \tfrac{1}{3}h \left(A_1 + \sqrt{A_1 A_2} + A_2\right),$$

here $\qquad A_1 =$ area of upper surface of frustum

$\qquad A_2 = \quad$,, $\quad$ lower $\qquad$,, $\qquad$,,

$\qquad h =$ the altitude.

WORKED EXAMPLE. The height of a frustum of a pyramid is 1·5 ft and the areas of its ends are 4·5 sq. ft. and 12·5 sq. ft. Find its volume

$$V = \tfrac{1}{3}h \,(A_1 + \sqrt{A_1 A_2} + A_2)$$
$$= \tfrac{1}{3} \times 1{\cdot}5\,(4{\cdot}5 + \sqrt{4{\cdot}5 \times 12{\cdot}5} + 12{\cdot}5)$$
$$= 0{\cdot}5\,(4{\cdot}5 + 7{\cdot}5 + 12{\cdot}5)$$
$$= 0{\cdot}5 \times 24{\cdot}5$$
$$= 12{\cdot}25 \text{ cu. ft.}$$

Special case: **Frustum of a cone.**

Method 1. Fig. 111 represents the frustum of a right cone. It is contained between two parallel planes AB and CD.

Its volume is obviously,

    Volume of cone OCD

         − volume of cone OAB.

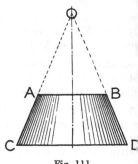

Fig. 111

Method 2. Use the general rule given above:

$$V = \tfrac{1}{3}h \,(A_1 + \sqrt{A_1 A_2} + A_2).$$

Exercises 67

1. What volume of water will fill a glass tumbler of height 5 in., th diameters at top and bottom being $3\tfrac{1}{2}$ in. and $2\tfrac{1}{2}$ in. respectively?

2. Calculate the weight of the stone cap shown in Fig. 112 if the stone weighs 156 lb. per cubic foot.   (*N.C.*)

3. Fig. 113 is a vertical section of a portion of a concrete floor and column. The column is circular and the cap is a frustum of a cone.

Calculate, to the nearest lb., the weight of concrete required to form the indicated portion of the column and cap if the density is 150 lb. per cubic foot. Ignore the plaster and floor finish.   (*N.C.*)

147. Sphere.

A sphere is a solid swept out by the rotation of a semi-circle about its diameter as axis.

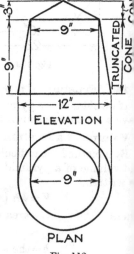

Fig. 112

Thus if we rotate the semi-circle shown at (a) in Fig. 114 about XY .he axis of rotation) it will sweep out "a solid sphere of revolution".

The sphere is a compact solid bounded by a curved surface which is everywhere equidistant from a point within it called the *centre*. 'he distance from the centre to any point on the surface is called .e *radius*.

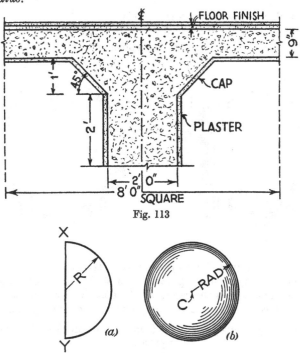

Fig. 113

Fig. 114

Any section of a sphere by a plane is a circle. This can easily be shown r cutting a wooden ball. If the section plane passes through the centre the sphere, the circle formed is called a *great circle of the sphere.* Such section would divide the sphere into two *hemispheres.*

148. Area of surface of sphere (Fig. 115).

Let r = radius, d = diameter.

It can be shown by means of a proof too difficult to introduce at this stage that if we suppose a cylinder to have the same diamete[r] as a sphere, and a height equal to that diameter,

Whole surface of sphere = curved surface of cylinder
$$= 2\pi r \times 2r$$
$$= \mathbf{4\pi r^2}, \quad \text{or} \quad \mathbf{\pi d^2}.$$

149. Volume of a sphere (Fig. 116).

Suppose a very small portion of the surface of a sphere to be [a] regular polygon. If its base area be considered sufficiently small w[e] may regard this base as a plane surface. If from every angular poin[t]

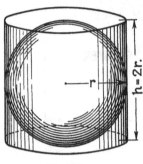

Fig. 115

Fig. 116

in this base we draw a line to the centre of the sphere we shall hav[e] a pyramid. The apex of the pyramid lies at the centre of the spher[e] and its altitude equals the radius of the sphere. Imagine the whol[e] sphere to be composed of a very large number of such pyramids.

Volume of sphere = $\frac{1}{3}$ (radius) × (sum of the areas of the bases o[f] the pyramids)

$$= \tfrac{1}{3}r \times 4\pi r^2 \quad \text{(for surface of sphere} = 4\pi r^2)$$

$$= \mathbf{\frac{4}{3}\pi r^3}, \quad \text{or} \quad \mathbf{\frac{\pi d^3}{6}}. \quad \left(\text{Note } \frac{\pi}{6} = 0{\cdot}5236.\right)$$

EXAMPLE 1. Find the volume of a sphere having a diameter of 10 in

$$\text{Volume} = \frac{\pi d^3}{6} = 0{\cdot}5236 d^3$$
$$= 0{\cdot}5236 \times 10^3$$
$$= 523{\cdot}6 \text{ cu. in.}$$

EXAMPLE 2. Find the surface area of a sphere, the volume of which is c.c.

Let us first find the diameter.

$$0 \cdot 5236 d^3 = 10,$$

$$\therefore \quad d = \sqrt[3]{\frac{10}{0 \cdot 5236}},$$

$$\therefore \quad \log d = \tfrac{1}{3} \,(\log 10 - \log 0 \cdot 5236)$$
$$= \tfrac{1}{3} \,(1 - \bar{1} \cdot 7190)$$
$$= 0 \cdot 4270.$$

$$\therefore \quad d = 2 \cdot 673 \text{ cm.}$$

$$\text{Surface area} = \pi d^2 = \pi \times 2 \cdot 673^2.$$

Let

$$\text{Area} = a,$$

$$\therefore \quad \log a = 0 \cdot 4972 + 2\,(0 \cdot 4270)$$
$$= 1 \cdot 3512.$$

$$\therefore \quad a = 22 \cdot 45 \text{ (sq. cm.).}$$

Exercises 68

$$\left(log\ \pi = 0 \cdot 4972, \quad log\ \frac{\pi}{4} = \bar{1} \cdot 8951, \quad log\ \frac{\pi}{6} = \bar{1} \cdot 7190. \right)$$

1. A stone sphere is 3 ft. in diameter, the stone weighing 120 lb. per bic foot. Find the weight of the sphere.

2. The circumference of the great circle of a sphere is 12 ft. What is s volume?

3. If two balls of metal 6 in. and 8 in. in diameter respectively are elted down to form another sphere, what will be its diameter and lume?

4. 80 solid spherical iron castings, 5 in. in diameter, are to be cast, p.c. extra metal being allowed for runner metal and wastage.
A cylindrical ladle 24 in. deep contains the necessary metal when $\frac{5}{8}$ full. Calculate the internal diameter of the ladle to 4 significant figures.

5. Find the volume of a hollow sphere, the inside and outside diaeters of which are 1·5 in. and 6 in.

6. If the area of a great circle of a sphere is 101·3 sq. in., what is its dius and what is the surface area of the sphere?

7. Graph to show the relation between the volume and diameter of a sphere. Complete the following table:

Diameter	0·3	0·5	0·7	0·9	1·1	1·2
Diameter³	0·027					
Volume	0·014					

(a) Plot a graph *Volume* v. *Diameter*. It is a curved graph.

(b) Plot a graph *Volume* v. *Diameter³*. It is a straight line. Find its slope.

SIMILARITY

150. Similar triangles.

Definition. Similar triangles are equi-angular.

Thus in Fig. 117 the triangles OAB and Oab are similar if

$$\angle O = \angle O,\ \angle A = \angle a,\ \angle B = \angle b.$$

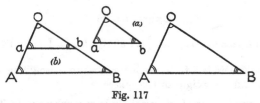

Fig. 117

To examine the relationship between the sides of similar triangles.

EXERCISE 1. Carefully draw two similar triangles as in Fig. 117 (a). Measure the sides and by substitution verify the following:

$$\frac{OA}{Oa} = \frac{OB}{Ob} = \frac{AB}{ab},$$

i.e. show that *in similar triangles corresponding pairs of lines are in the same ratio*.

EXERCISE 2. As in Fig. 117 (b) draw any triangle OAB. Draw ab parallel to AB.

AB and ab are parallels and OA is a transversal, whence

$$\angle Oab = \angle OAB \text{ (corresponding angles)}.$$

Similarly $\angle Oba = \angle OBA$ „   „

In the triangles Oab, OAB,

$$\left.\begin{array}{l} \angle Oab = \angle OAB \\ \angle Oba = \angle OBA \\ \angle AOB \text{ is common} \end{array}\right\} \qquad \therefore \quad \text{the triangles are similar.}$$

Measure, substitute and verify the following:

(1) $\dfrac{OA}{Oa} = \dfrac{OB}{Ob} = \dfrac{AB}{ab}$.

(2) $\dfrac{OA}{AB} = \dfrac{Oa}{ab}$; $\dfrac{AB}{BO} = \dfrac{ab}{bO}$; $\dfrac{BO}{OA} = \dfrac{bO}{Oa}$.

Conclusions.

(1) **In similar triangles the ratios of the pairs of corresponding sides are equal.**

(2) **In similar triangles the pairs of sides about equal angles are in the same ratio.**

(3) **When a line is drawn parallel to the base of a triangle, the smaller triangle thus produced is similar to the first triangle.**

151. To draw similar triangles by radial projection.

Method. See Fig. 118. Draw any $\triangle ABC$. Choose any point P outside

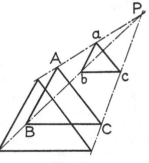

. From P draw *radials* through the angular points of the triangle. By means of set-square and ruler draw similar triangles having sides parallel to those of ABC and having their vertices on the radials.

Similar triangles in Fig. 118: ABC, abc; Pab and PAB; Pbc and PBC; Pac and PAC, etc.

(a) *Ratios of pairs of corresponding sides are equal:*

$$\frac{AB}{ab} = \frac{AC}{ac} = \frac{BC}{bc} = \frac{PB}{Pb} = \frac{PA}{Pa}, \text{ etc.}$$

Fig. 118

(b) *Sides about the equal angles are in the same proportion:*

$$\frac{ab}{bc} = \frac{AB}{BC}, \quad \text{and} \quad \frac{bc}{ca} = \frac{BC}{CA}, \quad \text{and} \quad \frac{ca}{ab} = \frac{CA}{AB}.$$

152. Similar Polygons.

Definition. Similar polygons are (1) *equi-angular and* (2) *have corresponding pairs of sides about the equal angles in the same ratio.*

Fig. 119 (*a*). By radial projection we have obtained two similar polygons *ABCDE* and *abcde*.

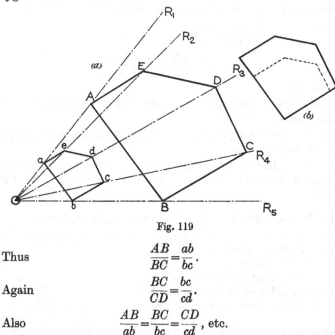

Fig. 119

Thus
$$\frac{AB}{BC}=\frac{ab}{bc}.$$

Again
$$\frac{BC}{CD}=\frac{bc}{cd}.$$

Also
$$\frac{AB}{ab}=\frac{BC}{bc}=\frac{CD}{cd}, \text{ etc.}$$

Fig. 119 (*b*). Equi-angular polygons are not necessarily similar polygons. Thus at (*b*) the polygons are equi-angular but dissimilar— for they have not the same shape, i.e. the sides about the equal angles are not in the same ratio. In defining similar polygons, therefore, we must refer to sides as well as angles.

153. Other similar figures.

All circles are similar to one another; so also are all squares, all equilateral triangles, all regular pentagons. In the solids all spheres are similar, so are all cubes. When the ratio of length to diameter is the same in both we have similar cylinders.

154. The areas of similar figures.

(1) If side of a square $= 2$ in., its area $= 2^2$ sq. in.

       ,,   ,, $= 3$ in.,   ,, $= 3^2$,,

$$\therefore \quad \frac{\text{Area of large square}}{\text{Area of small square}} = \frac{3^2}{2^2} = \frac{\text{Side}^2}{\text{side}^2}.$$

(2) If side of equilateral triangle $= a$ in., its area $= 0\!\cdot\!433a^2$ sq. in.

      ,,   ,, $= A$ in.,   ,, $= 0\!\cdot\!433A^2$,,

$$\therefore \quad \frac{\text{Area of large triangle}}{\text{Area of small triangle}} = \frac{0\!\cdot\!433A^2}{0\!\cdot\!433a^2} = \frac{A^2}{a^2} = \frac{\text{Side}^2}{\text{side}^2}.$$

(3) If diameter of a circle $= d$ in., its area $= \dfrac{\pi}{4} d^2$.

      ,,   ,, $= D$ in.,   ,, $= \dfrac{\pi}{4} D^2$.

$$\therefore \quad \frac{\text{Area of small circle}}{\text{Area of large circle}} = \frac{\dfrac{\pi}{4} D^2}{\dfrac{\pi}{4} d^2} = \frac{D^2}{d^2}.$$

Thus the area of a square or equilateral triangle is proportional to the square of its side; that of a circle to the square of its diameter.

The areas of similar figures are proportional to the squares of corresponding linear dimensions.

EXAMPLE. Given an irregular hexagon on a base of 2 in. it is required to draw another hexagon similar to the first but having an area four times as great. (See Fig. 120.)

      Let $S =$ side of large hexagon,

      and $s =$,,   small   ,,   .

$$\therefore \quad \frac{\text{Area of large hexagon}}{\text{Area of small hexagon}} = \frac{S^2}{s^2} = \frac{4}{1},$$

but $s = 2$ in.,

$$\therefore \quad \frac{S^2}{2^2} = \frac{4}{1},$$

$$\therefore \quad S^2 = 16,$$

$$\therefore \quad S = 4 \text{ in.}$$

Exercises 69

1. Draw an equilateral triangle of 2 in. side and another of double the area of the first.

2. If the area of an equilateral triangle on a base of 10 in. is 43·3 sq. in. find the area of a similar triangle on a base of 2 in.

3. The area of a regular hexagon on a base of 10 in. is 259·8 sq. in. Find the area of a similar polygon on a base of 2 in.

4. The shadow of a poplar tree is 140 ft. long while the shadow of a yard stick at the same time and place is 43 in. long. Find the height of the tree.

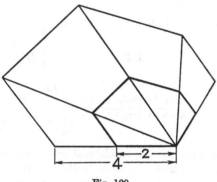

Fig. 120

5. How many circles of 1 in. diameter will equal the area of one of 2 in. diameter?

6. What proportion does the area of a steam pipe of 8 in. diameter bear to the cross-sectional area of a cylinder of 40 in. diameter?

7. There are two maps, one to a scale of 2 in. = 1 mile, and the other to a scale of $\frac{1}{2}$ in. = 1 mile. The area of an estate on the first map is 1·46 sq. in. What is the area of the same estate on the second?

155. Similar solids.

1. The **surfaces** of similar solids are proportional to the *squares* of corresponding edges or of any corresponding lines that may be drawn on them.

2. The **volumes** of similar solids are proportional to the *cubes* of

ɔrresponding edges or of any corresponding lines which may be
rawn on them.

EXAMPLE. A cube has an edge of 12 in. Another cube has double the
olume of the first. What is the length of its edge?

$$\frac{\text{vol.}}{\text{Vol.}} = \frac{(\text{edge})^3}{(\text{Edge})^3}.$$

$$\therefore \quad \frac{1}{2} = \frac{12^3}{E^3},$$

$$\therefore \quad E^3 = 2 \times 12^3 = 3456,$$

$$\therefore \quad E = \sqrt[3]{3456},$$

$$\therefore \quad \log E = \tfrac{1}{3} \log 3456$$
$$= 1{\cdot}1795,$$

$$\therefore \quad E = 15{\cdot}12 \text{ in.}$$

Exercises 70

1. If an iron sphere weighs 6 lb. and is $3\frac{1}{2}$ in. diameter, find the
ameter of another sphere of the same material weighing 20 lb.

2. The ratio of the volumes of two spheres being $1 : 4{\cdot}035$, find the
ameter of the larger if that of the smaller is $2{\cdot}105$ in.

3. A right cone weighing 216 lb. is 24 in. high. It is cut by a plane
ɑrallel to the base and 12 in. from the top. What is the weight of the
ɔmaining portion, or frustum, of the cone?

4. A block of wood has been cut to scale to copy the exact shape of a
ɪece of marble 21 in. by 16 in. by 11 in. The wood weighs 17 lb. and its
ɑaterial weighs 48 lb. per cubic foot. Find the dimensions of the wood.

CHAPTER XV

CURVED GRAPHS

156. We have seen that the graphs of equations of the form $y = mx + c$ are straight lines. If, however, the equation is of the form $y = mx^2 + c$ the resulting graph is a *curve*.

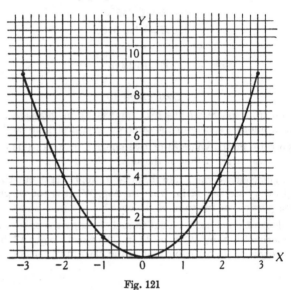

Fig. 121

EXAMPLE 1. Draw the graph of $y = x^2$ from $x = -3$ to $x = +3$. Tabulating values:

x	-3	-2	-1	0	1	2	3
y	9	4	1	0	1	4	9

Plotting we get the curve in Fig. 121. This curve is called a parabola

EXAMPLE 2. Draw the graph of $y = x^2 - 2x - 3$ from $x = -2$ to $x = +4$.

Tabulating values:

x	-2	-1	0	1	2	3	4
x^2	4	1	0	1	4	9	16
$-2x-3$	1	-1	-3	-5	-7	-9	-11
y	5	0	-3	-4	-3	0	5

Plotting we get the graph Fig. 122.

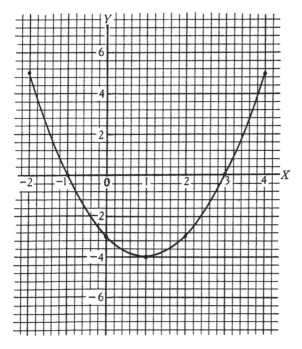

Fig. 122

Exercises 71

1. Draw the graph of the equation
$$y = x^2 + 2x - 8 \text{ from } x = -5 \text{ to } x = +3.$$

2. Draw the graph of the equation
$$y = x^2 - 4x - 5 \text{ from } x = -2 \text{ to } x = +6.$$

3. On the same axes draw the graph of $y = x^2$ from $x = -3$ to $x = +4$ and the graph of $y = x + 6$. What are the co-ordinates of the points where the straight line cuts the curve?

4. Draw the graph of $y = x^2 - \dfrac{3x}{2} - 1$ from $x = -1$ to $x = +3$.

5. Draw the graph $y = x^2$ from -1 to $+3$, also the straight line $y = 1\frac{1}{2}x + 1$. What are the co-ordinates of the points where the straight line cuts the curve?

CHAPTER XVI

QUADRATIC EQUATIONS

157. A quadratic equation is one which contains the square but no higher power of the unknown quantity.

The simplest form of a quadratic equation contains the unknown quantity expressed in the second power only—for example $x^2 = 16$. This equation may be solved by simply extracting the square root, therefore $x = 4$. But as 16 may be produced by multiplying -4 by -4, the square root of 16 may be either $+4$ or -4, written ± 4, therefore $x = \pm 4$, and $+4$ and -4 are called the **roots of the equation.**

EXAMPLE 1. Solve the equation $5x^2 = 180$.

$$x^2 = 36,$$
$$\therefore \quad x = \pm 6.$$

EXAMPLE 2. $(x-4)^2 = 1\frac{9}{16}$.

Since
$$(x-4)^2 = \frac{25}{16},$$
$$\therefore \quad x-4 = \pm \frac{5}{4}.$$

Taking $+\frac{5}{4}$,
$$x = \frac{5}{4} + 4$$
$$= \frac{21}{4}.$$

Taking $-\frac{5}{4}$,
$$x = 4 - \frac{5}{4}$$
$$= \frac{11}{4}.$$
$$\therefore \quad x = \frac{21}{4} \text{ or } \frac{11}{4}.$$

Note. Every quadratic equation has two roots.

158. To solve a quadratic equation when the unknown quantity is expressed both in its second and first powers. This may be done by three methods.

Method 1. By factors.

EXAMPLE 1. Solve the equation $x^2 = 3x + 4$.

Step 1. *Bring all terms to left hand side of equation.* Thus
$$x^2 - 3x - 4 = 0.$$

Step 2. *Factorise.* Thus $(x-4)(x+1)=0$.

If the product of two or more factors $=0$, each in turn may be $=0$.

$$\therefore \quad x-4=0,$$

whence
$$x=4,$$

also
$$x+1=0.$$

$$\therefore \quad x=-1.$$

$$\therefore \quad x=4 \text{ or } -1.$$

EXAMPLE 2. Solve the equation $x^2-8x=9$.

Step 1.
$$x^2-8x-9=0.$$

Step 2.
$$(x-9)(x+1)=0.$$

$$\therefore \quad x-9=0,$$

whence
$$x=9,$$

or
$$x+1=0,$$

whence
$$x=-1.$$

$$\therefore \quad x=9 \text{ or } -1.$$

Method 2. By completing the square.

EXAMPLE 1. Solve the equation $x^2=12x+13$.

Step 1. Arrange the equation with the unknown quantity on the left, and the known on the right.

Then
$$x^2-12x=13.$$

Step 2. The left side of the equation is next made a perfect square. This can always be done by adding to it *the square of half the coefficient of x*.

The same addition must, of course, be made to *both* sides of the equation.

The equation now becomes

$$x^2-12x+(6)^2=13+(6)^2,$$

that is
$$(x-6)^2=13+36=(7)^2.$$

$$\therefore \quad x-6=\pm 7,$$

whence
$$x=13 \text{ or } -1.$$

Note. *Before adding the square of half the coefficient of x, care must be taken to arrange that the coefficient of x^2 is unity.*

EXAMPLE 2. Solve $2x^2+x-6=0$.

$$2x^2+x=6.$$

(Divide throughout by the coefficient of x^2.)

$$\therefore \quad x^2 + \tfrac{1}{2}x = 3.$$

Completing square $x^2 + \tfrac{1}{2}x + (\tfrac{1}{4})^2 = 3 + (\tfrac{1}{4})^2,$

hat is $$(x + \tfrac{1}{4})^2 = 3 + \tfrac{1}{16} = \tfrac{49}{16} = (\tfrac{7}{4})^2,$$

$$\therefore \quad (x + \tfrac{1}{4}) = \pm \tfrac{7}{4},$$

whence $$x = \tfrac{3}{2} \text{ or } -2.$$

Summarising.

1. Arrange the terms, so that only the numerical is on the right hand side of the equation.

2. Divide across by the coefficient of x^2.

3. Add the square of half the coefficient of x to both sides of the equation.

4. Take the square root of both sides.

5. Solve the resulting simple equations.

Method 3. Solution by formula.

The general form of a quadratic equation is

$$ax^2 \pm bx \pm c = 0.$$

If methods 1 and 2 present any difficulty in application, the following formula may be used. Arrange all terms on left hand side as above.

Then $$x = \frac{\mp b \pm \sqrt{b^2 \mp 4ac}}{2a}.$$

Note. *In the formula, the sign of the coefficient of x, i.e. of b, is always opposite to what it is in the equation and the sign of $4ac$ is always opposite to the sign of c in the equation.*

EXAMPLE 1. Solve $2x^2 - 7x = 15$.

Arranging terms $$2x^2 - 7x - 15 = 0.$$

By formula $$x = \frac{7 \pm \sqrt{7^2 + 4.15.2}}{2.2}$$

$$= \frac{7 \pm \sqrt{49 + 120}}{4}$$

$$= \frac{7 \pm \sqrt{169}}{4}$$

$$= \frac{7 \pm 13}{4}$$

$$= 5 \text{ or } -\tfrac{3}{2}.$$

$a = 2$
$b = -7$
$c = -15$

Formula used

$$x = \frac{+b \pm \sqrt{b^2 + 4ac}}{2a}$$

EXAMPLE 2. Solve $2x^2 + 11x + 12 = 0$.

$$x = \frac{-11 \pm \sqrt{11^2 - 4.2.12}}{2.2}$$

$$= \frac{-11 \pm \sqrt{121 - 96}}{4}$$

$$= \frac{-11 \pm \sqrt{25}}{4}$$

$$= \frac{-11 \pm 5}{4}$$

$$= \tfrac{3}{2} \text{ or } -4.$$

$a = 2$
$b = 11$
$c = 12$

159. In addition to the three algebraic methods of solving a quadratic equation there is the graphic method. Referring to Fig. 12 in para. 156, the curve is the graph of the quadratic equation $x^2 - 2x - 3 = 0$. Where this curve cuts the OX axis, viz. where $x =$ and $x = -1$ gives the roots of the equation because these two values satisfy the equation $x^2 - 2x - 3 = 0$.

160. Another graphic method is indicated in Questions 3 and 5 of Exercises 71. Here in Question 3 the equation is $x^2 - x - 6 = 0$ because since

$$y = x^2 \quad \text{and} \quad y = x + 6,$$
$$\therefore \quad x^2 = x + 6,$$
$$\therefore \quad x^2 - x - 6 = 0.$$

We therefore put $x^2 = x + 6 = y$, plot the curve $y = x^2$ and the straight line $y = x + 6$. The curve and straight line intersect at points $x = 3$ and $x = -2$. These then are the roots of the equation $x^2 - x - 6 = 0$.

Exercises 72

Solve the following by any of the foregoing methods:

1. $x^2 - 4x = 12$.
2. $x^2 + 2x - 24 = 0$.
3. $x^2 + 4x = 21$.
4. $x^2 - 6x = 27$.
5. $x^2 - 9x = -20$.
6. $x^2 + 10x + 21 = 0$.
7. $x^2 - 3x = 40$.
8. $2x^2 - 7x = 15$.
9. $3x^2 - 20x + 12 = 0$.
10. $10x^2 + 23x = 5$.
11. $x^2 - 9x + 20 = 0$.
12. $x^2 - 7x - 8 = 0$.
13. $x^2 + 3x - 18 = 0$.
14. $x^2 + 11x + 28 = 0$.

15. $x^2 - 6x + 5 = 0.$

16. $x^2 - 9x + 8 = 0.$

17. $x^2 + 30 = 11x.$

18. $3x^2 - 20x - 7 = 0.$

19. $2x^2 - x - 6 = 0.$

20. $3x^2 - 21x - 24 = 0.$

21. $5x^2 + 6x - 8 = 0.$

22. $4x^2 - 13x - 12 = 0.$

23. $3x^2 - 22x = 16.$

24. $2x^2 - 21x = 98.$

25. $8x^2 + 22x = 21.$

26. $2x^2 - 19 \cdot 5x = 5.$

27. $\dfrac{x^2}{8} + 2\tfrac{11}{12}x = 2.$

28. $\dfrac{x-1}{3} = \dfrac{5}{x+1}.$

29. $\dfrac{x-1}{x+3} = \dfrac{2x-5}{x}.$

30. $\dfrac{4x+1}{3x+2} = \dfrac{x-1}{x-2}.$

31. $2 \cdot 5x^2 + 2x = 0 \cdot 5.$

32. $0 \cdot 6x^2 - 4x - 1 \cdot 4 = 0.$

Exercises 73

1. Divide 17 into two parts, so that one-third the product of the parts is 20.

2. Find a number whose square divided by 7 is equal to 3 times the number diminished by 14.

3. Find a number, which when squared and increased by 1, will equal 15 times the number, decreased by 25.

4. The sum of the reciprocals of two consecutive whole numbers is $\frac{7}{12}$. Find the numbers.

5. A wire 20 in. long is bent so as to form a rectangle. If the area of this rectangle is 22·75 sq. in., find its dimensions.

6. A field is 3 times as long as it is broad, and contains 1 acre 3 roods 20 perches. Find the cost of fencing it at $4\tfrac{1}{2}d.$ per yard.

7. The distance from P to Q is 45 miles. A man cycles 9 miles per hour faster than he can walk and so does the journey in $11\tfrac{1}{4}$ hours less. Find his rate of walking.

8. The hypotenuse of a right-angled triangle is 13 ft. The perpendicular is 7 ft. longer than the base. Find the base and perpendicular.

9. The hypotenuse of a right-angled triangle is 2·5 in., and its perimeter is 6 in. Find the base and perpendicular.

10. The length of the side of a square courtyard is 30 yd. It contains a square grass plot, surrounded by a path of uniform width. If the cost of paving the path is £3. 12s. 0d., at the rate of $1\tfrac{1}{2}d.$ per square yard, find its width.

11. I spend £13. 10s. 0d. on drills. Had they cost 1s. 6d. per doze
less, two dozen more could have been purchased for the same outlay
Find the cost per dozen.

12. There are 360 square panes of glass in a house. Had the side of th
pane been 1 in. longer, 110 fewer panes would have been required. Wha
is the size of the pane?

13. The total ages of a class of boys are 459 years. If two new boy
are admitted aged respectively $8\frac{1}{2}$ and $9\frac{1}{2}$, the average age of the class i
lowered 3 months. How many boys were originally in the class, and wha
was their average age?

14. The bed of a reservoir with sloping banks measures 30 yd. b
40 yd., and when the water is 12 ft. deep the surface of the water measure
38 yd. by 44 yd. Find the depth of the water when the area of th
surface is 1800 sq. yd.   (*C. and G*

15. The hypotenuse of a right-angled triangle is 61 in., and the sur
of the three sides is 132 in. Find the base and perpendicular.

16. A man spends £15 a year on coal. The price is increased by 16s. 8d
a ton, so he finds that he will have to use $1\frac{1}{2}$ tons less a year, if his outla
is to be the same as before. What was the original consumption of coal

17. The sum of the squares of two numbers is 145 and their produc
72. Find them.

18. A certain number of yards of belting cost £4. 10s. 0d. Had th
price been 2d. a yard less, 6 more yards could have been bought. Fin
how many yards were bought, and the price per yard.

19. A square and a rectangle are equal in area. The side of th
square is 3 in. longer than the breadth of the rectangle. If the length c
the rectangle were diminished by 3 in. and its breadth increased by 1 in
the area would remain unaltered. Find the dimensions of the rectangle

20. A rectangle is 9 ft. long. On its ends as diameters are describe
semi-circles. The total area thus formed is 101·5 sq. ft. Find the breadt
of the rectangle. $(\pi = 3\frac{1}{7}.)$

Exercises 74

Solve the following equations by graphs:

1. $x^2 - x - 6 = 0$.   **2.** $x^2 - x - 12 = 0$.

3. $x^2 + 3x + 12 = 0$.   **4.** $x^2 - 10x + 16 = 0$.

5. $x^2 - 5x + 4 = 0$.   **6.** $x^2 - 0·5x - 1·5 = 0$.

7. $2x^2 - x - 1 = 0$.   **8.** $2x^2 - x - 3 = 0$.

9. $2x^2 - 5x - 12 = 0$.   **10.** $0·5x^2 - 3·5x + 3 = 0$.

CHAPTER XVII

TRIGONOMETRY

161. Trigonometry literally means *three cornered measurement.*
t deals primarily with the relations between the sides and angles of
triangle.

PRACTICAL EXAMPLE 1. (1) Construct a right-angled triangle, the
ase of which is 3 units, and perpendicular 6 units. (Make your drawing
s large as the paper will allow.) Letter your drawing as in Fig. 123,

$$\frac{AB}{BC} = \frac{6}{3}.$$

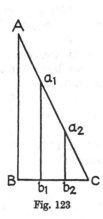

Fig. 123

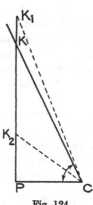

Fig. 124

(2) Draw perpendiculars a_1b_1, a_2b_2 as in Fig. 123.
Measure these perpendiculars and their corresponding bases b_1C, and
C as accurately as possible. Tabulate the result of the ratio of each
erpendicular to its base as shown.

$\frac{AB}{BC}$	$\frac{a_1b_1}{b_1C}$	$\frac{a_2b_2}{b_2C}$

(3) Now by means of a protractor or tracing paper, construct an
ngle the same size as the angle at C (Fig. 123). Take any point P

(Fig. 124) in the base line and at this point erect a perpendicular cutting the other leg of the angle C at K.

Measure PK and PC and compare the ratio $\dfrac{PK}{PC}$ with the values of the ratios in table in (2).

(4) Make the angle $PCK_1 >$ the angle PCK (Fig. 124).

Now because $PK_1 > PK$

$$\therefore \quad \frac{PK_1}{PC} > \frac{PK}{PC}.$$

(5) Make the angle $PCK_2 < PCK$ (Fig. 124).

Because $\qquad\qquad PK_2 < PK,$

$$\therefore \quad \frac{PK_2}{PC} < \frac{PK}{PC}.$$

Conclusion. *We therefore see that the ratio of the perpendicular to the base depends upon the size of the angle opposite the perpendicular. This ratio is called the* **Tangent Ratio** *of the angle C.*

Definition. The tangent of an angle is the ratio of the perpendicular opposite that angle to the base intercepted between the foot of the perpendicular and the angle.

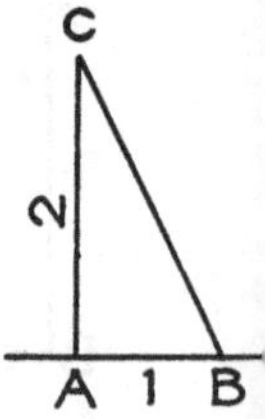

PRACTICAL EXAMPLE 2. Construct an angle whose tangent is 2.

Draw a base line, and mark off 1 unit AB (Fig. 125), at A erect a perpendicular AC of 2 units. Join BC.

$$\therefore \quad ABC \text{ is the angle required,}$$

for $\qquad \tan \angle B = \dfrac{\text{perpendicular}}{\text{base}} = \dfrac{AC}{AB} = 2.$

Fig. 125

Exercises 75

1. Using your protractor construct the following angles, 20°, 5 30°, 35°, 40°, 45°, 50°, 55°, 60°. Mark off 5 units on the base line each, and erect perpendiculars (as in Fig. 123). Measure these pendiculars in terms of the unit chosen, and calculate the value of th tangent of each angle. (Make your drawings as large as the paper w allow.)

2. Construct a graph of the results obtained in Question 1. The axis to represent degrees, and the OY axis, tangent values.

3. Construct an angle whose tangent is 0·4. Measure it in degrees.

4. Construct and measure in degrees the angle whose tangent is 2.

5. A ladder leaning against a wall makes an angle of 72° with the pavement. The top of the ladder is 20 ft. above the pavement. What distance is the foot of the ladder away from the wall? (Fig. 126.)

6. The tread of a stairway is 10 in., and its rise 7 in. At what angle does the stairway slope? (Fig. 126.)

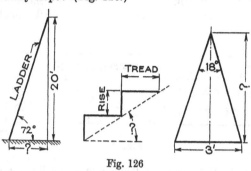

Fig. 126

7. On squared paper plot the point (5, 8), and find at what angle the line joining the point to the origin meets the *OX* axis.

8. The vertical angle of an isosceles triangle is 18°. Its base is 3 ft. Find the vertical height. (Fig. 126.)

9. Construct a right-angled triangle *ABC*. The angle at *C* is 90°, angle *ABC* = 13°, side *BC* = 10 units. Measure *CA* and calculate tan 13°.

10. In a right-angled triangle *ABC*, *AC* is 20 in., angles *ACB* and *BAC* are 90° and 55° respectively. Find the side *BC* and the area of the triangle.

Definition. *The angle of elevation of an object is the angle contained between the line joining the point of observation with the object, and the horizontal line passing through the point of observation. The angle BEO is the angle of elevation of object at O* (Fig. 127).

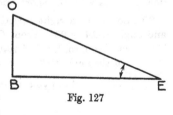

Fig. 127

11. The angle of elevation of the top of a wireless mast, observed from a distance of 500 ft., is 30°. If the mast stands on a hill 260 ft. high, what is the length of the mast?

12. A stick casts a shadow of 5 ft. when the elevation of the sun i 60°. What is the length of the stick?

13. The sides of a roof slope at an angle of 35° with the horizontal The ridge of the house is 12 ft. above the level of the end wall. Find th width of the house. (Fig. 128.)

14. A bath is 42 ft. long. It is 12 ft. deep at one end and 6 ft. deep a the other. At what angle does the bottom of the bath slope? (Fig. 128.

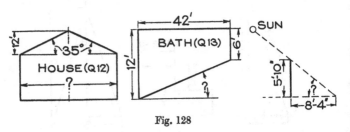

Fig. 128

15. The base of an isosceles triangle is 20 in. long and its vertica height 4·04 in. Find the base angles to the nearest degree.

16. A man whose height is 5 ft. 10 in. casts a shadow 8 ft. 4 in. Wha is the elevation of the sun to the nearest degree? (Fig. 128.)

17. Two vertical posts are 20 ft. apart, their heights being 22 and 30 f respectively. Find the inclination to the horizontal of a tight-rope joining their tops.

18. The co-ordinates of two points are (3, −1), and (−3, −4). A what angle does the line joining these two points meet the *OX* axis?

19. What is the length of the longest piece of wood that would li along the bottom of a drawer 2 ft. long and 10 in. broad. What angl does it make with the short side of the drawer?

20. Construct a right-angled triangle *ADB*. The angle *DAB* is 74 and angle *BDA* is 90°. From *D* drop a perpendicular *DO* on *AB*. A is 3 units. From tables, find value of tan 74°, thence calculate *DO*. Fin the size of the angle *BDO*, thence find length of *BO*. Compare the rati $\frac{DO}{AO}, \frac{BO}{DO}, \frac{BD}{AD}$. (Make your drawing as large as possible.)

SINES AND COSINES

162. It has been shown that the ratio of the perpendicular to the base in a right-angled triangle is the same, no matter at what point in the base a perpendicular is erected. Likewise it may be shown that the ratio of the perpendicular to the hypotenuse will remain constant for the same angle.

EXAMPLE. Draw a base line AB 3 units long, and at one extremity erect a perpendicular AC 4 units long. (Fig. 129.) Measure BC. It will be found to be 5 units long. The ratio of AC to CB is $\frac{4}{5}$. Take any other points a_1, a_2, a_3 in the base and erect perpendiculars intersecting CB in c_1, c_2 and c_3 respectively.

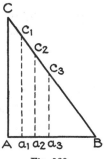

Fig. 129

Measure the length of each perpendicular, also the lengths c_1B, c_2B, c_3B. Now determine the values of the ratios

$$\frac{a_1c_1}{c_1B}, \quad \frac{a_2c_2}{c_2B}, \quad \frac{a_3c_3}{c_3B},$$

and compare these values with the value $\frac{4}{5}$.

This ratio of the perpendicular to the hypotenuse is called the **sine ratio**, and $\dfrac{AB}{CB}$ is said to be the sine of the angle B, written sin B.

It may likewise be shown that $\dfrac{AB}{BC}$ is equal to

$$\frac{a_1B}{Bc_1}=\frac{a_2B}{Bc_2}=\frac{a_3B}{Bc_3}, \text{ viz. } \frac{3}{5}.$$

The ratio of the base to the hypotenuse is called the **cosine ratio**, and $\dfrac{AB}{BC}$ is said to be the cosine of the angle B, written cos B.

163. Now consider the angle at C (Fig. 129). Turn your book round until CA becomes horizontal. Then AB is perpendicular and C is still the hypotenuse.

The sine of the angle at C (sin C) is equal to the ratio $\dfrac{AB}{BC}$. But in the foregoing this was shown to be the cosine ratio of the angle at A (cos B).

$$\therefore \quad \cos B = \sin C.$$

In like manner it may be shown that $\sin B = \cos C$. Now $\angle B + \angle C = 90°$, i.e. they are complementary angles. From this is derived a very important relation, namely, **the sine of an angle is equal to the cosine of its complement,**

i.e. $$\sin A = \cos (90 - A).$$

Test this statement by referring to your tables of sines and cosines.

164. In any right-angled triangle

$$(\text{base})^2 + (\text{perpendicular})^2 = (\text{hypotenuse})^2$$

i.e. $$a^2 + b^2 = c^2 \text{ (Fig. 130)}.$$

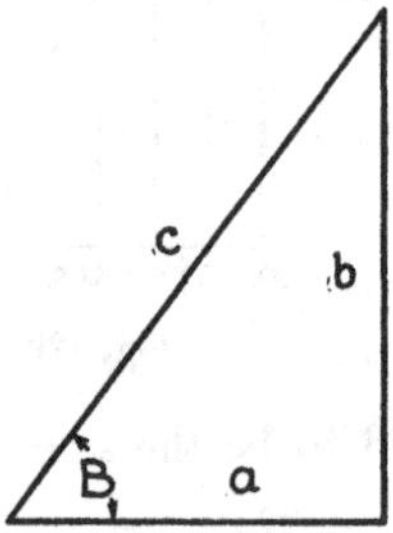

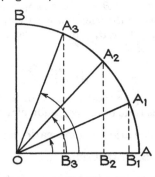

Fig. 130

Fig. 131

Dividing this equation across by c^2, we get

$$\frac{a^2}{c^2} + \frac{b^2}{c^2} = 1.$$

But $$\frac{a}{c} = \cos B \quad \text{and} \quad \frac{b}{c} = \sin B$$

$$\therefore \quad \cos^2 B + \sin^2 B = 1.$$

165. An angle is the measurement of the amount of turning of a straight line revolving round a fixed point, one end of the line being on the fixed point.

In Fig. 131 the line OA revolves around the fixed point O. Three positions OA_1, OA_2, OA_3 in the path of revolution are shown.

The angle AOA_1 measures the movement from A to A_1.
    " AOA_2 "   "   "   " A to A_2.
    " AOA_3 "   "   "   " A to A_3.

Imagine that at A there is a hinged line which hangs vertically s the line OA revolves. This telescopic line appears as A_1B_1, A_2B_2, A_3B_3 as the revolving line OA reaches the positions OA_1, OA_2 and OA_3 respectively. Consider the ratios of the angles AOA_1, AOA_2, AOA_3:

$$\sin AOA_1 = \frac{A_1B_1}{OA_1} = \frac{\text{perpendicular}}{\text{hypotenuse}}.$$

$$\sin AOA_2 = \frac{A_2B_2}{OA_2} = \frac{\text{perpendicular}}{\text{hypotenuse}}.$$

$$\sin AOA_3 = \frac{A_3B_3}{OA_3} = \frac{\text{perpendicular}}{\text{hypotenuse}}.$$

But $$OA_3 = OA_2 = OA_1,$$

and $$A_3B_3 > A_2B_2 > A_1B_1.$$

$$\therefore \quad \frac{A_3B_3}{OA_3} > \frac{A_2B_2}{OA_2} > \frac{A_1B_1}{OA_1},$$

that is, $\sin AOA_1$ is less than $\sin AOA_2$ and $\sin AOA_2$ is less than $\sin AOA_3$. We have thus established that **the greater the angle the greater the value of its sine.**

166. Consider next the cosine values of the same angles:

$$\cos AOA_1 = \frac{OB_1}{OA_1} = \frac{\text{base}}{\text{hypotenuse}}.$$

$$\cos AOA_2 = \frac{OB_2}{OA_2} = \frac{\text{base}}{\text{hypotenuse}}.$$

$$\cos AOA_3 = \frac{OB_3}{OA_3} = \frac{\text{base}}{\text{hypotenuse}}.$$

But $$OA_1 = OA_2 = OA_3,$$

and $$OB_1 > OB_2 > OB_3.$$

$$\therefore \quad \frac{OB_1}{OA_1} > \frac{OB_2}{OA_2} > \frac{OB_3}{OA_3}.$$

Therefore cos AOA_1 is greater than cos AOA_2, and cos AOA_2 is greater than AOA_3. Another very important truth has been established, namely, **the greater the angle the less the value of its cosine.**

167. Considering the tangent values of the same angles, it can be seen from Fig. 131 that as the angle increases in size, so also does its tangent value. **The greater the angle, the greater the value of its tangent.**

EXAMPLE 1. Construct an angle whose cosine is 0·5.

Then $\cos A = 0·5 = \frac{1}{2}$,

which means we have to construct a right-angled triangle whose base is 1 unit and hypotenuse 2 units.

Draw a base line AX (Fig. 132). Mark off a distance $AB = 1$ unit.

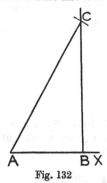

Fig. 132

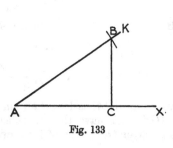

Fig. 133

At B erect a perpendicular of unlimited length.

With A as centre and a radius $= 2$ units, describe an arc, cutting the perpendicular at the point C.   Join AC.

Then CAB is the required angle.

$$\cos CAB = \frac{AB}{AC} = \frac{1}{2}.$$

Measure this angle with your protractor and compare it with the angle given in the tables whose cosine $= \frac{1}{2}$.

EXAMPLE 2. The hypotenuse of a right-angled triangle is 8 units and the base angle is 34°. Find the lengths of base and perpendicular.

Draw a base line AX (Fig. 133). At point A make an angle of 34°, i.e. angle XAK.

On AK mark off a length AB of 8 units.

From B drop perpendicular BC. Measure BC and CA in terms of the units in AB.

To check these measurements,

$$\cos 34° = \frac{AC}{AB} = \frac{AC}{8}.$$

∴ $8 \cos 34° = AC.$ (*Look up* $\cos 34°$ *in the tables.*)

.e. $8 \times 0.8290 = AC.$

       $6.632 = AC.$

$$\sin 34° = \frac{BC}{AB} = \frac{BC}{8}.$$

∴ $8 \sin 34° = BC.$ (*Look up* $\sin 34°$ *in the tables.*)

.e. $8 \times 0.5592 = BC.$

       $4.4736 = BC.$

Exercises 76

1. Construct an angle whose sine is 0.8. Measure it with your protractor, and compare with the tables.

2. A boy requires 120 ft. of string to fly his kite. The elevation of the ite is 50°. How high is it above the ground?

3. Referring to Fig. 131 write down sin 90°, cos 90°, tan 90°.

4. A flagpole is 42 ft. high, and is stayed by a wire rope. A man 6 ft. all stands 15 ft. away from the pole, and finds that his head just touches he wire stay. Calculate (*a*) the distance from the top of his head to the op of the pole, (*b*) the angle the stay makes with the ground.

5. A pole 26 ft. high breaks at a point 6 ft. from the ground and falls ver. What angle does the top make with the ground?

6. A ladder makes an angle of 58° with the ground. How far up the adder will a string 10 ft. long have to be fastened, so that, hanging ertically, it just reaches the ground?

7. A man walks 2 miles along a road, which rises at an angle of 2°. y how much has he increased his vertical height above sea level?

8. The sides of a triangle are in the proportion of 3, 4 and 5. What is he size of the smallest angle?

9. A man whose height is 6 ft. 3 in. walks down a hill which slopes at n angle of 5°. How far has he gone when his head is level with the top the hill?

10. A pendulum is 30 in. long and swings through an angle of 20°. How far does the "bob" rise?

11. A circle has a radius of 13 in. Find the size of the angle at the centre subtended by a chord 24 in. long.

12. The 30° angle of a thin set-square can be pushed into a hollow cylindrical tube, with the shorter of the two edges lying along the tube. If this part of the edge inside the tube measures 8 cm., find the diameter of the tube.

13. An aerial pole has a stay which makes an angle of 50° with the ground. On being lengthened 12 ft. it is found that the inclination is now 40°. Find the height of the pole to the nearest foot.

14. In a triangle ABC, the angle BCA is 65°, and angle BAC is 25°. Find the lengths of AB and BC. The length of AC is 125 units.

15. The area of the section of a square prism is 6·25 sq. in. What is the area of a section making an angle of 60° with this section?

16. $ABCD$ is a trapezium in which the sides AD and BC are parallel. The side BA makes an angle of 52° with AD. CD is perpendicular to the parallel sides. $BC = 8·5$, $BA = 6·7$. Find CD and AD.

17. A man whose eye is 5 ft. above the ground sees the tops of two posts in a straight line. He knows that their heights are respectively 13 ft. and 21 ft., and that they are 30 yd. apart. How far is he from the nearer post?

18. A person standing on the bank of a river observes that the elevation of a tree on the opposite bank is 60°. He walks backward 40 ft. and then finds that its elevation is 30°. Find the height of the tree and the breadth of the river.

19. A ladder 29 ft. long rests against a wall making with it an angle of 30°. If the foot of the ladder is pulled $5\frac{1}{2}$ ft. farther from the wall, how far up the wall will the top of the ladder reach?

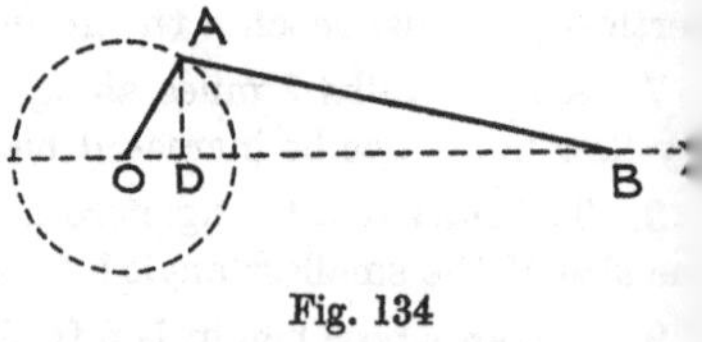

20. In Fig. 134 OA represents a crank 18 in. long which revolves in the plane of the paper about O. AB is a rod 6 ft. long jointed at A to the crank. As the crank revolves the end B slides along a pathway OX. At the moment when the angle AB is 5°, how far is B from O? Answer correct to nearest inch.

Fig. 134

168. The six trigonometrical ratios.

In addition to the three ratios already dealt with there are three others. These three additional ratios are the *cosecant*, the *secant* and the *cotangent*. These are respectively the reciprocals of the sine, the cosine, and the tangent. Tabulating the six ratios we have:

$$\sin A = \frac{\text{perpendicular}}{\text{hypotenuse}}, \quad \operatorname{cosec} A = \frac{\text{hypotenuse}}{\text{perpendicular}}.$$

$$\cos A = \frac{\text{base}}{\text{hypotenuse}}, \quad \sec A = \frac{\text{hypotenuse}}{\text{base}}.$$

$$\tan A = \frac{\text{perpendicular}}{\text{base}}, \quad \cot A = \frac{\text{base}}{\text{perpendicular}}.$$

169. The tangent of an angle is equal to the cotangent of its complement.

In Fig. 135 $\tan B = \dfrac{AC}{CB}$, and $\cot A = \dfrac{AC}{CB}$.

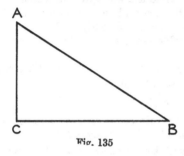

Fig. 135

CHAPTER XVIII

GEOMETRY OF THE CIRCLE

170. A **circle** is a plane figure bounded by a curved line (th*e circumference*) which is everywhere equi-distant from a fixed poin*t* (the *centre*). See Fig. 136.

A **radius** is a straight line drawn from the centre to the circumference.

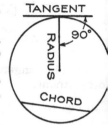

A **diameter** is a straight line passing through the centre and terminated at both ends by the circumference.

A **chord** is a straight line joining any two points in the circumference. It divides the circle into two **segments.**

A **tangent** to a circle is a straight line which meets the circumference and though produced indefinitely does not cut it. See Fig. 136.

Fig. 136

Concentric circles are described from the same centre but hav*e* different radii.

A **sector** is bounded by two radii and that part of the circum*-* ference cut off between them.

171. *Proposition* 9.

(a) **The perpendicular drawn from the centre of a circle to** *a* **chord bisects the chord,** *or conversely*;

(b) **The straight line drawn from the centre to the mid**-*point* **point of the chord is perpendicular to the chord.**

Proof. (a) See Fig. 137. *CD* is any chord of a circle of centre *O*. Let *OP* be the perpendicular from *O* to *CD*.

To prove that *P* bisects *CD*.

In the triangles *OCP, ODP,*

$$\begin{cases} OC = OD \text{ (radii of same circle),} \\ OP = OP, \\ \angle OPC = \angle OPD \text{ (by construction).} \end{cases}$$

Therefore triangles are congruent, and *CP = PD.*

(b) Let *P* be the mid-point of *CD*.

Fig. 137

To prove that OP is perpendicular to CD.

In the triangles OCP, ODP,

$$\begin{cases} OC = OD \text{ (radii of same circle)}, \\ OP = OP, \\ CP = PD \text{ (by data)}. \end{cases}$$

Therefore triangles are congruent, and $\angle OPC = \angle OPD$. But these are adjacent and supplementary. Therefore each is a right angle and OP is perpendicular to CD.

Useful riders.

1. The perpendicular bisector of a chord of a circle passes through the centre.

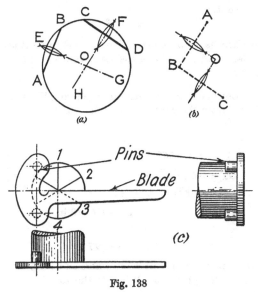

Fig. 138

See Fig. 138 (a) where two non-parallel chords AB and CD are bisected at right angles by EG and FH respectively. These bisectors pass through the centre O.

In Fig. 138 (b) we are given three points A, B and C and require to find the centre of the circle which flows through them. Join AB and BC. Bisect each at right angles. The bisectors intersect at O, the required centre.

A Practical Application. Sheet Metal Centre-Finder.

In Fig. 138(c) we show a type of **centre-finder** frequently used for locating the centres of round shafts. It consists of a blade and stock cut out of sheet metal, the edge of the blade being exactly mid-way between, and at right angles to, the line joining the centres of the two inserted pins (which project from the under-side of the stock). On placing the square on the end of a shaft, with the pins pressed up against its curved surface, lines may be scribed along the edge of the blade. Two or more such lines (1, 2, 3 on the figure) will pass through the centre, for each is the perpendicular bisector of a chord.

2. Equal chords in a circle are equi-distant from the centre.

See Fig. 137. Let r = radius of circle, $2s$ = length of chord, t = distance of chord from centre.

Then $r^2 = t^2 + s^2.$

If another chord is drawn equal in length to the first it is clear that in the second case r and s will have the same values and t must have the same value also.

EXAMPLE. Calculate the length of a chord CD which lies at a distance of 3 in. from the centre of a circle of 5 in. radius.

See Fig. 139.

$$CT^2 = OC^2 - OT^2.$$
$$\therefore \quad CT = \sqrt{OC^2 - OT^2}$$
$$= \sqrt{25 - 9} = \sqrt{16} = 4 \text{ in.}$$

Thus $CD = 8$ in.

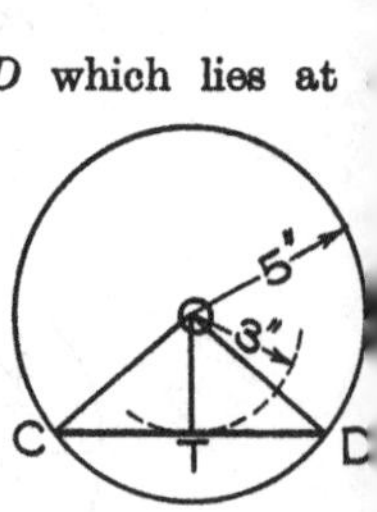

Fig. 139

Exercises 77

1. Take three points A, B and C, anywhere on your paper and describe a circle to flow through them. When is the solution of this problem impossible?

2. In a circle of $1\frac{1}{2}$ in. radius, draw two chords 2 in. and $2\frac{1}{2}$ in. long respectively. Calculate the distance of each from the centre and check your result by measurement.

3. CD is a chord 3 in. long in a circle of centre O and 2 in. radius. Find the area of triangle OCD.

4. A horse-shoe arching is in the form of an arc comprising three-quarters of the circumference of a circle standing on the chord. The diameter of the circle is 10 ft. Calculate the width of the road. (*Hint: Revise para. 95 and check by drawing to scale.*)

5. Two circles of diameters 74 in. and 40 in. respectively have a common chord 24 in. long. Find the distance between their centres. Note that these circles intersect each other, the common chord joining their points of intersection.

172. Angles in circles.

The chord AB in Fig. 140 divides the circle into two segments—the *major segment ADB* and the *minor segment ACB*. If O is the centre, the $\angle AOB$ is called *the angle at the centre*. Note that there are two angles AOB; one is the obtuse $\angle AOB$ and is said to be *subtended at the centre by the arc ACB*, the other is the reflex $\angle AOB$ which is said to be *subtended at the centre by the arc ADB*. Again, if D is any point on the major arc, $\angle ADB$ is said to be *the angle in the segment ADB*. Similarly $\angle ACB$ is said to be *the angle in the segment ACB*. Further $\angle ADB$ is said to *stand on the arc ACB*, whilst $\angle ACB$ *stands on the arc ADB*.

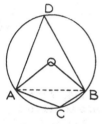

Fig. 140

173. *Proposition* 10. **An angle at the centre of a circle is double the angle at the circumference standing on the same arc.**

Let AEB (Fig. 141 (*a*)) be an arc of a circle of which O is the centre, and let C be any point on its circumference. Join AC, CB.

Then ACB is an angle at the circumference standing on the arc AEB.

Join AO, OB.

We have to prove that $\angle AOB=$ twice $\angle ACB$.

Join CO and produce to any point D.

Then $\qquad \angle DOA = \angle OAC + \angle OCA$ (para. 76).

But $\qquad\qquad OA = OC$.

$\therefore \quad \angle OCA = \angle OAC$ (para. 82).

$\therefore \quad \angle DOA =$ twice $\angle OCA$.

Similarly $\angle DOB =$ twice $\angle OCB.$

$\therefore$ $\angle DOA + \angle DOB =$ twice $(\angle OCA + \angle OCB).$

$\therefore$ $\angle AOB =$ twice $\angle ACB.$

In Fig. 141(b) the $\angle AOB$, at the centre, subtends an arc AE which is greater than the arc of a semi-circle. In this case $\angle AC$ is obtuse, whilst $\angle AOB$ is reflex. The proposition holds good fo all cases, whence $\angle ACB = \frac{1}{2}\angle AOB.$

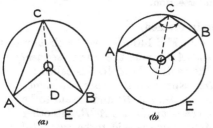

Fig. 141

174. *Proposition* 11. **Angles in the same segment of a circl are equal to one another.**

Let ACB, AC^1B, AC^2B be angles in the same segment of the circl whose centre is O (Fig. 142).

To prove that these angles are equal.

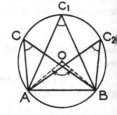

Proof. By Prop. 10.

$\angle AOB = 2\angle ACB.$

$\angle AOB = 2\angle AC_1B.$

$\angle AOB = 2\angle AC_2B,$ etc.

$\therefore$ $2\angle ACB = 2\angle AC_1B = 2\angle AC_2B.$

$\therefore$ $\angle ACB = \angle AC_1B = \angle AC_2B.$

Fig. 142

175. *Proposition* 12. **The angle in a semi–circle is a righ angle.**

Let APB be a semi-circle, of which O is the centre, and let P be an point on the circumference (Fig. 143).

Then, by Prop. 10, $\angle AOB = 2\angle APB.$

But $\angle AOB$ is a straight angle,

whence $\angle APB$ is a right angle.

176. A quadrilateral is said to be **cyclic** when a circle can be described to flow through its four angular points.

Proposition 13. **The sum of two opposite angles of a cyclic quadrilateral is two right angles.**

Let *ABCD* be a cyclic quadrilateral and *O* the centre of the ircle (Fig. 144).

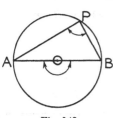

Fig. 143

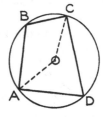

Fig. 144

To prove that

$$\angle A + \angle C = 180°, \quad \text{and} \quad \angle B + \angle D = 180°.$$

Proof. Join *OA*, *OC*.

By Prop. 10 $\angle CBA = \frac{1}{2}$ reflex $\angle COA$.

,, $\angle CDA = \frac{1}{2}$ obtuse $\angle COA$.

∴   Adding, $\angle CBA + \angle CDA = \frac{1}{2}$ (sum of angles at *O*).

∴ $\angle B + \angle D = \frac{1}{2}$ (360°) = 2 right angles.

Similarly, $\angle A + \angle C = 2$ right angles.

Conversely, if two opposite angles of a quadrilateral are supplementary, the quadrilateral is cyclic.

Exercises 78

1. Draw an isosceles triangle *ABC* having a vertical angle *ABC* = 70°, nd a base *AC* = $2\frac{1}{2}$ in. With centre *B*, and radius *BA*, describe a circle. bove the chord *AC* mark a series of points on the circumference as ·, *M*, *N*, etc. Measure the angles *ALC*, *AMC*, *ANC*, etc. Compare hem with the angle *ABC*, and with one another.

2. Two adjacent sides of a cyclic quadrilateral inscribed within a ircle of $2\frac{1}{2}$ in. radius subtend angles of 70° and 40° at the centre. The emaining two sides are equal. Draw the quadrilateral, measure its ngles and verify by calculation.

3. Draw an isosceles triangle and on one of the equal sides describ
a circle. Why does it flow through the mid-point of the base?

4. Prove that the opposite angles of a cyclic quadrilateral are supple
mentary.

In a quadrilateral *ABCD* of this type the angles *CAD, BDA, BDC* ar
15°, 65°, 35° respectively. Find the angles of the quadrilateral an
construct it if *AC* = 1·7 in. Measure length of *BD*.

5. Fig. 145 represents a plot of building land. Draw it to a scale c
1 in. = 12 ft. Measure *CD* and verify by calculation.

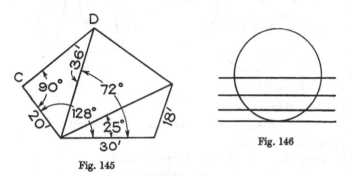

Fig. 145   Fig. 146

TANGENCY

177. A **secant** is a straight line which cuts the circumference of
circle in two points. Suppose a secant moves downwards from th
centre as shown in Fig. 146. The points in which it cuts the cir
cumference gradually approach each other until they completel
coincide. Then the straight line touches the circle at two coinciden
points or, in other words, *one* point. In this position the straigh
line is called a **tangent**.

178. *Proposition* 14. **A tangent to a circle is perpendicula
to a radius drawn to a point of contact.**

GRAPHICAL EXERCISES.

1. To draw a tangent to a circle at any point on its circumference.

Method. See Fig. 147 (*a*). Suppose *P* is the point on the circumference
Join *P* to the centre *O*. Then *PO* is a radius. At *P* erect a perpendicula
to *PO*.

2. To draw a tangent to a circle from a point without it.

Method. See Fig. 147 (*b*). Suppose the circle is of $1\frac{1}{4}$ in. radius and $P = 3\frac{1}{4}$ in. Draw *PO* and bisect it. Describe a semi-circle upon it, cutting circumference in *V*. Join *PV*, *VO*. Similarly the tangent *PW* may be drawn.

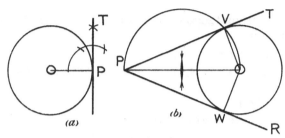

Fig. 147

Analysis. (*a*) Which are the points of contact between tangents and circle?

(*b*) Which are the angles between the tangents and the radii drawn to the points of contact?

(*c*) State the magnitude of these angles.

(*d*) Prove that *PVOW* is a cyclic quadrilateral.

3. To draw two tangents to a circle of $1\frac{1}{4}$ in. radius to include an angle of 40° between them.

Method. See Fig. 148.

$\angle X + \angle Y = 180°$; $\angle P + \angle C = 180°$.

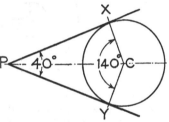

Fig. 148

Thus we commence by drawing radii *CX*, *CY*, having an angle of $180 - 40 = 140°$ between them. At *X* and *Y* erect perpendiculars meeting at *P*.

Exercises 79

1. About a circle of 2 in. diameter, circumscribe an equilateral triangle.

2. About a circle of $2\frac{1}{2}$ in. diameter, circumscribe a quadrilateral having two opposite angles right angles and a third angle 60°. What is the magnitude of the fourth angle?

3. The inscribed circle of a triangle is the circle tangential to all three sides. If such a circle has a diameter of 2·3 in., and two angles of the triangle are 50° and 70° respectively, draw both circle and triangle and measure the length of the longest side. What is the area of the triangle?

179. A **common tangent** to two circles is a straight line which is tangential to both.

GRAPHICAL EXERCISES.

1. To draw an exterior common tangent to two circles which are not tangential.

Method. (Fig. 149.)

(*a*) Draw the line of centres *OC*.

(*b*) Bisect it and describe a semi-circle on it.

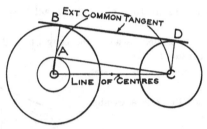

Fig. 149

(*c*) With centre *O*, and radius equal to the difference between the radii of the given circles, describe a circle.

(*d*) Draw *AC* tangential to this circle. Join *OA* and produce it to meet the larger circle in *B*.

(*e*) Through *C* draw *CD* parallel to *OA*.

(*f*) Join *BD*. It is the required tangent.

2. To draw an interior common tangent to two circles which are not tangential.

Method. (Fig. 150.)

(*a*) Draw the line of centres *OC*.

(*b*) Bisect it and describe a semi-circle on it.

(*c*) With centre *O*, and radius equal to the sum of the radii of the given circles, describe a circle. It cuts the semi-circle in *F*.

Join *OF*. Then *D* is a point of contact.

(*d*) Through *C* draw *CE* parallel to *OF*.

(*e*) Join *DE*. It is the required tangent.

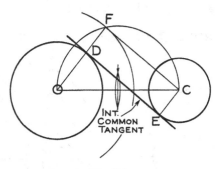

Fig. 150

Exercises 80

1. How many common tangents can be drawn in each of the following cases: (*a*) when the circles intersect, (*b*) when they touch externally, (*c*) when they touch internally?

2. The distance between the centres of two circles of radii $2\frac{1}{4}$ in. and in. respectively is $2\frac{1}{2}$ in. Draw an exterior common tangent. Measure its length and verify by calculation.

3. In Fig. 151 *EF* is a chord of a circle and *FG* a tangent to the circle at *F*. Find, geometrically, the centre of the circle, and measure the angles in its larger segment.

4. *ABC* is a triangle having *BC* = 3 in., *CA* = 2·8 in., *AB* = 2·6 in. Draw accurately a circle to pass through *A* and to touch *BC* at its mid-point.   (*C and G.L.I.*)

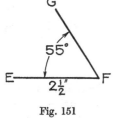

Fig. 151

5. In a triangle whose sides are 7, 9, 11 cm. long respectively, inscribe a circle. Measure the lengths of the tangents from the angular points of the triangle to this circle.

180. *Proposition* 15. **The angle made by a tangent to a circle with the chord drawn from the point of contact is equal to the angle in the alternate segment of the circle.**

Let AB be a tangent to the circle (in Fig. 152) at D. Let DE be a chord. Draw a diameter DF, and join FE. Then by Prop. 1 $\angle DFE = \angle DGE$ or any other angle in the major segment.

Let DCE be any angle in the minor segment.

To prove that

$$\angle BDE = \angle DFE, \quad \text{and} \quad \angle ADE = \angle DCE.$$

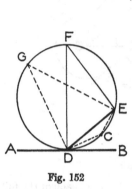

Fig. 152

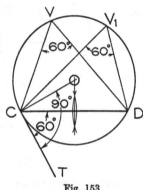

Fig. 153

Proof. (i) Since $DCEF$ is a semi-circle, $\angle FED$ is a right angle.

$$\therefore \quad \angle DFE = 90° - \angle EDF.$$

But $\qquad\qquad \angle BDF$ is a right angle.

$$\therefore \quad \angle BDE = 90° - \angle EDF.$$

$$\therefore \quad \angle BDE = \angle DFE.$$

(ii) $FDCE$ is a cyclic quadrilateral.

$$\therefore \quad \angle DCE = 180° - \angle DFE \quad \text{(para. 176)}.$$
$$= 180° - \angle EDB.$$
$$= \angle ADE.$$

GRAPHICAL EXERCISE. To obtain a segment of a circle to contain any given angle—say 60°.

Method. See Fig. 153. Suppose we are given the chord CD, 2 in. long.

(1) Bisect CD at right angles.

(2) Draw CT at 60° to CD.

(3) Draw CO perpendicular to CT and meeting the perpendicular bisector of CD in O.

(4) With centre O, and radius OC, describe a circle, in the major segment of which draw any angles such as CVD, CV_1D.

(5) Angle DCT is the angle between the tangent and the chord. Angles CVD, CV_1D are angles in the alternate segment.

Exercises 81

1. In a circle of 3 in. diameter, inscribe a triangle having angles of 45°, 65°, 70°.

2. Draw a triangle ABC in which the base $AB = 2$ in. vertical angle $ACB = 60°$, side $BC = 2\frac{1}{8}$ in. How many solutions are there?

3. Draw a triangle CDE being given the base $CD = 2\frac{1}{2}$ in., vertical angle $CED = 55°$, altitude $= 1\frac{3}{4}$ in. How many solutions are there?

4. Draw to a scale of one-tenth full-size a quadrilateral $ABCD$ as in Fig. 154. (*Hint: Start with AC.*)

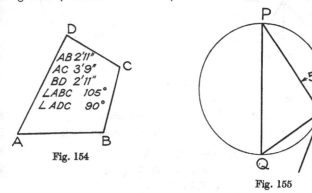

AB 2′11″
AC 3′9″
BD 2′11″
∠ABC 105°
∠ADC 90°

Fig. 154

Fig. 155

5. Fig. 155 (which is not drawn to scale) shows a circle of 3·4 in. diameter. A chord PA is drawn making an angle $PAT = 55°$ with the tangent AT. The diameter PQ is drawn and Q is joined to A. What are the sizes of the angles PQA, APQ, QAP? Calculate the lengths of AP, PQ.   (*R.A.F. Entrance.*)

181. *Proposition* 16. **When two or more chords of a circle intersect at a point within it, the rectangles contained by their segments are equal.**

INTERESTING EXPERIMENTAL VERIFICATION.

(1) Describe any circle. It is well to choose a bold radius—say 2 in

(2) Choose any point O (not the centre) and through it draw an number of chords as GH, FE, CD, AB in Fig. 156.

(3) Measure the segments (or parts) of each chord and calculate th product of the segments of each. Set down your results thus:

Chord AB. Product of segments $AO \times OB =$
   ,, CD.   ,,   ,, $CO \times OD =$
   ,, FE.   ,,   ,, $FO \times OE =$
   ,, GH.   ,,   ,, $GO \times OH =$

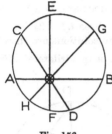

Fig. 156

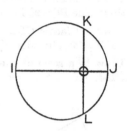

Fig. 157

IMPORTANT SPECIAL CASE (Fig. 157).

(1) Let the chord IJ be a diameter.

(2) Through a point O on IJ (*not* the centre) erect a perpendicular KI

(3) The circle being symmetrical about a diameter, K is the *image* L and $OK = OL$.

$$\therefore \quad OK \times OL = OK^2.$$

(4) $$IO \times OJ = OK^2 \quad (Prop.\ 16).$$

$$\therefore \quad \sqrt{IO \times OJ} = OK.$$

Application. To obtain a square equal in area to a given rectangl (Fig. 158).

Method. Let $ABCD$ be the given rectangle. With centre B, and radiu BC, describe an arc cutting AB produced in X. Bisect AX and describ a semi-circle on it. Produce BC to meet the semi-circle in G. On B construct the square $BGFE$.

Then
$$AB \times BC = BG^2.$$
$$\therefore \quad \sqrt{AB \times BC} = BG.$$

We call BG the mean proportional between AB and BC. We may
xpress it thus:

r

$$AB : BG :: BG : BC \quad \text{\}\textit{product of means equals}$$
$$BC : BG :: BG : AB \quad \text{\}\textit{product of extremes.}$$

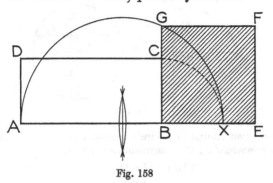

Fig. 158

Exercises 82

1. In a circle of $3\frac{1}{2}$ in. diameter, draw a chord $2\frac{3}{4}$ in. long. Draw a
iameter at right angles to the chord. Measure the segments of the
hord and of the diameter. Find the product of the segments of each
nd compare results.

2. Draw a rectangle 3 in. by 2 in. Reduce it to a square of equal
rea and write down the length of its side.

3. Find by construction the mean proportional between two lines
·8 in. and 2·3 in. long respectively. Check your result arithmetically.

4. Reduce a regular pentagon of 2 in. base to a square of equal area.
Hint: *First reduce the pentagon to a triangle of equal area.*)

5. P is a point 6 in. distant from the centre of a circle of 10 in. radius.
tate a construction for drawing through P the chord of least length.
rove the truth of your construction and find the length of the chord of
ast length.

If through P a chord APB is drawn such that AP is 5 in. long, find by
lculation the length of PB.   (*R.A.F. Entrance.*)

182. WORKED EXAMPLE. A segmental arch has a rise of 20 ft., and a
an of 82 ft.

(*a*) Find the diameter of the circle of which the arc is a portion.

(b) Find the angle at the centre of the circle subtended by the arc.

(c) Find the length of the soffit, i.e. the curve of the arch.

(U.E.I.

Answer: See Fig. 159.

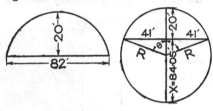

Fig. 159

(a) When two chords of a circle intersect at a point within it th
rectangles contained by their segments are equal.

Then $20 \times x = 41 \times 41.$

$$\therefore \quad x = \frac{41 \times 41}{20} = 84 \cdot 05.$$

$$\therefore \quad \text{Diameter} = 20 + 84 \cdot 05 = 104 \cdot 05 \text{ ft.}$$

(b) $\text{Sin} \angle \theta = \frac{41}{52 \cdot 025} = \cdot 7885.$

$$\therefore \quad \theta = 52°.$$

$$\therefore \quad \text{Whole angle at centre} = 104°.$$

(c)   Length of soffit $= \frac{104}{360} \times \pi d$

$$= \frac{104}{360} \times \frac{22}{7} \times \frac{104 \cdot 05}{1}$$

$$= 94 \cdot 5 \text{ ft.}$$

Exercises 83. (The Circle)

1. A plano-convex lens is $\frac{3}{16}$ in. thick, and 5 in. diameter. Find th
radius of the sphere from which it was cut. (See
Fig. 160.)

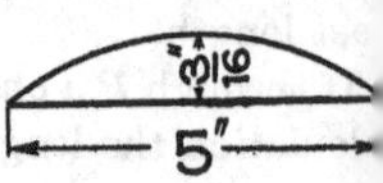

2. The radius of a circular archway is 20 ft. If
its span is 30 ft., find its rise.

Fig. 160

3. (a) Draw a rectangle 7 cm. by 3 cm., reduce it to a square of equa
area. Measure the side of the square.

(b) Find $\sqrt{21}$, firstly by drawing intersecting chords within a circle, secondly by the theorem of Pythagoras.

Compare results.

4. Draw a triangle XYZ so that $XY = 3 \cdot 7$ in., $YZ = 3 \cdot 9$ in., $ZX = 4 \cdot 1$ in. Find the centre O of a circle which contains the three points XYZ on its circumference. Join O to Z and Y thus forming the triangle OZY. Measure the angles ZXY and ZOY and state them as a ratio.

5. Prove that an exterior angle of a cyclic quadrilateral is equal to the interior opposite angle.

6. A rectangle is inscribed within a circle. Prove that its diagonals intersect at the centre of the circle.

CHAPTER XIX

LOCI

183. PRACTICAL EXERCISES.

1. Given a point O it is required to find the path of a point free to move so that it remains in the same plane as O and keeps $1\frac{1}{2}$ in. from it.

Method. See Fig. 161 (*a*). O is the fixed point. With centre O, and radius $1\frac{1}{2}$ in., describe a circle. This is the *locus* or path of the moving point.

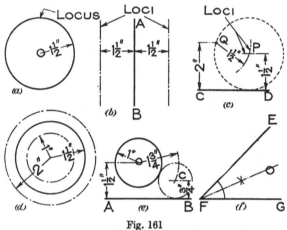

Fig. 161

2. Given a straight line AB it is required to find the path of a point which moves so as to be $1\frac{1}{2}$ in. from AB.

Method. See Fig. 161 (*b*). AB is the fixed straight line. On each side of it draw parallel straight lines $1\frac{1}{2}$ in. from it. These are the paths or *loci* required.

3. Given a point Q and a straight line CD 2 in. from it. Required to find a point P which shall be $1\frac{1}{2}$ in. from Q and from CD.

Method. (*a*) With centre Q, and radius $1\frac{1}{2}$ in., describe an arc. This line contains all points $1\frac{1}{2}$ in. from Q. (*b*) Draw a line $1\frac{1}{2}$ in. from C.

nd parallel to it. All points in this line are $1\frac{1}{2}$ in. from CD. (c) P is the *intersection of the loci* and is thus $1\frac{1}{2}$ in. from both the given point Q and the straight line CD. (Fig. 161 (c).)

4. Given a circle of radius $1\frac{1}{2}$ in., it is required to obtain the paths of points which move so as always to be $\frac{1}{2}$ in. from the given circle.

Method. See Fig. 161 (d). With centre O and radius $1\frac{1}{2}$ in., describe the given circle. With the same centre and radii 1 in. and 2 in., describe "loci circles". All points on these circles are $\frac{1}{2}$ in. from the given circle.

5. Given a circle of centre O and radius 1 in., and a straight line AB $\frac{1}{2}$ in. from O, it is required to describe a circle of $\frac{3}{4}$ in. radius, which shall be tangential to both the given circle and the straight line.

Method. See Fig. 161 (e). (a) With centre O, and $1\frac{3}{4}$ in. radius, describe an arc. (b) Draw the locus of points $\frac{3}{4}$ in. from AB. (c) The intersection of the loci is C, a point $\frac{3}{4}$ in. from both circle and straight line, and thus the centre of the required circle.

6. Given an angle EFG, it is required to find the locus of the centres of circles tangential to the arms of this angle.

Method. See Fig. 161 (f). Bisect the angle. The bisector is the required locus. Take any point O on the bisector. From it drop perpendiculars to the arms of the angle. If you can prove the triangles to be congruent you have proved that the perpendiculars are equal.

Exercises 84

1. Draw a straight line AB 6 in. long. With centres A and B respectively describe circles of 4 in. radius. In how many points do the circles intersect? Mark two points.

2. Draw a straight line EF 3·5 in. long. Find a point which is 2·5 in. from E and F. How many such points are there?

3. Draw a straight line GH 3·6 in. long. Find a point I which is 2·7 in. from G, and 2·6 in. from H. How many such points are there?

4. Two battleships are anchored at a distance of 24 kilometres. The guns of each have an effective range of 14,000 metres. Draw a plan to a suitable scale showing the position of a target hit simultaneously by each ship.

5. A circle $2\frac{1}{4}$ in. diameter has its centre 2 in. from a given straight line. Describe a circle of 2 in. diameter which is tangential to both the given straight line and the circle.

The ellipse drawn with string and drawing pins (Fig. 162).

Take a piece of string or stout cotton and tie its ends together
Draw a straight line AB on your drawing paper and insert tw
drawing pins F and F_1 so that when the string is looped around then
it is slack. Pull the string taut with the point of a pencil and trac
out the path of the pencil point.

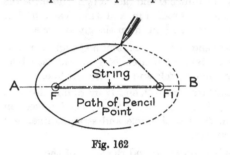

Fig. 162

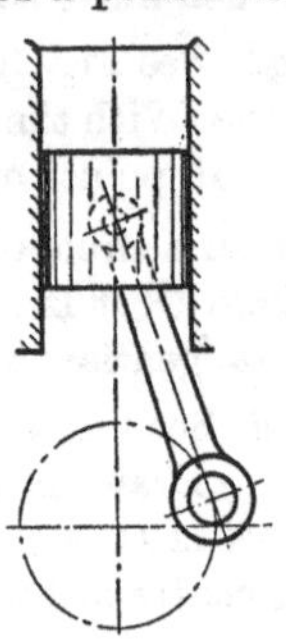

Fig. 163

The path or locus of the pencil point is an *ellipse* of which F and F
are called *foci*.

Exercises 85

1. Fig. 163 represents the piston and connecting rod of a petr
engine. What is the locus of the gudgeon pin and of the crank pin?

2. Draw two straight lines OA and OB at 60°. Draw a circle of 2 in
radius to touch both these lines. Then draw a smaller second circle t
touch both these lines and the first circle.   (*C. and G*

3. A steel plate $ABCD$ is quadrilateral shaped and has its side A
40 in. long, perpendicular to the side BC which is 80 in. long. The pe
pendicular distance of the corner D to the side BC is 30 in., and to th
side AB 40 in. Set the plate out to a scale of $\frac{1}{10}$th.

4. Construct the quadrilateral $ABCD$ from the following dat
$AB = 3$ in., $\angle ABC = 70°$, $\angle BAD = 100°$, $CA = CB$, and the perpendicul
distances of D from AB and BC are equal.   (*U.E.*

5. (*a*) If the base of a triangle is 2 in., and its altitude is $1\frac{1}{2}$ in., sta
clearly what is the locus of its vertex and draw the locus.

(*b*) If the vertical angle of a triangle is 30° and the base is 2 in., wh
is the locus of its vertex? Draw this locus also.

(c) Now construct accurately a triangle on a base of 2 in. whose altitude is $1\frac{1}{2}$ in., and vertical angle 30° showing all construction lines.

(N.B. The diagrams should be quite distinct from one another.)

(R.A.F.)

6. *AB* is a fixed rod. *AC* and *CX* are rods each 2 ft. long and hinged at *A* and *C*. *X* is a ring fixed to the end of *CX* and free to slide along *AB*. Find the locus of *C* as *X* slides along *AB*.

7. Find the locus of the mid-points of equal chords inscribed within a circle.

8. Find the locus of a point on the circumference of a circle when the circle rolls along a fixed straight line. (Note: *The resulting curve is a cycloid, useful to engineers and builders. The circle can be drawn on a piece of tracing paper and carefully rolled along a straight line.*)

9. Find the locus of a point on a straight line when the line is rolled round the circumference of a circle. (Note: *The locus is the involute of the circle. This curve is used as the profile of most modern toothed gears.*)

10. Draw a circle of 3 in. diameter and a straight line cutting it. Describe four circles each of $\frac{5}{8}$ in. radius, each tangential to the straight line and the circle.

11. Within a scalene triangle inscribe a circle. Its centre is known as the *in-centre* of the triangle.

12. Circumscribe a circle about any acute-angled scalene triangle. It will flow through the angular points or vertices. Its centre is known as the *circum-centre* of the triangle. Now find the circum-centre of (a) a right-angled triangle and (b) an obtuse-angled triangle.

TESTS

A

1. Find the L.C.M. of 7, 8, 15, 24.

2. If 50 yards of belting cost £6. 15s. 5d., find the cost of 39 yards

3. Solve
$$x - \frac{3}{5} - \frac{5\,(x-2)}{4} = \frac{3\,(x-\frac{1}{10})}{2}.$$

4. Evaluate $a^2c - 2b^2 + c^2$, if $a = 2$, $b = 3$, $c = 5$.

5. A rectangular room is 19 ft. long and 15 ft. wide. What is its area It is to be covered with linoleum 2 yd. wide at 6s. 7d. per sq. yd Assuming that there is no waste in cutting find the cost.

6. Construct a rhombus having its sides 3 in. and one angle 60° State (i) the magnitude of the remaining angles, (ii) the area of th figure in square inches.

B

1. Simplify $\qquad (1\frac{1}{2} + \frac{5}{6} \times \frac{9}{10}) \div (1\frac{1}{2} \times \frac{5}{6} - \frac{9}{10}).$

2. Fig. 164 shows two *diagonal scales* (a) to read inches, tenths an hundredths of an inch, (b) to read inches, eighths and sixty-fourths o

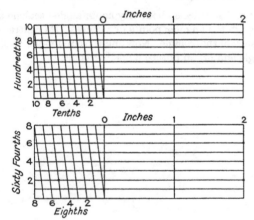

Fig. 164

n inch. (i) State the distances between the pairs of dots on each scale. (ii) Construct a diagonal scale of $1\frac{1}{2}$ in. = 1 yd. to show yards, eet and inches.

3. Solve $\dfrac{x-3}{4}+\dfrac{x-2}{5}=1+\dfrac{x-1}{40}$.

4. Evaluate $3x^2-2xyz+4z^2-\dfrac{5y^2}{2z^2}$,

$$x=2,\ y=3,\ 2z=5.$$

5. On the floor of a room is a carpet 15 ft. 7 in. long and 10 ft. 10 in. ide. There is an uncarpeted margin 18 in. wide all round. Find in quare feet the area not carpeted.   (R.S.A.)

6. Construct a right-angled triangle given that its height is 1·9 in. nd its hypotenuse 3·2 in. Measure its base, and its acute angles to ne nearest degree. Verify by calculation.

C

1. A bill of £13. 12s. 6d. was paid with 40 coins, some of them alf crowns and the remainder half sovereigns. Find the number f each kind.

2. Solve $2x+4y=16,\quad 6x=8y+3.$

3. Eight different operators took the following times, in seconds,) turn a wheel blank: 109, 112, 114, 119, 120, 117, 116, 113. Find ne average time.

4. Find the area of (i) a semi-circle of $\frac{3}{8}$ in. radius, (ii) a quadrant f a circle $\frac{5}{8}$ in. radius, (iii) a fillet of $\frac{3}{4}$ in. radius.

5. If $a=0$, $b=2$, $c=4$, $d=6$, find the value of

$$4\sqrt{2a+b^2+2dc-3}.$$

6. $ABCD$ is a trapezium AB and DC being parallel lines, $AB=5$in., $C=2\frac{1}{2}$ in., $CD=3\frac{1}{2}$ in., angle $D=90°$. Find AD, and the area of the gure.

D

1. The average of the cost of seven operations on a turned crank-aft is £2. 8s. 6d. The introduction of an eighth operation raises the verage by 1s. $9\frac{1}{2}d$. What was the cost of the eighth operation?

218 TESTS

2. A cyclist covers a certain distance in $2\frac{1}{2}$ hours, in returning hi
speed is 2 miles per hour less and he takes an hour longer. Fin
the distance and his speed on the outer journey.

3. The slant height of a cone is 55 ft. and the vertical height i
42 ft. Find its volume in cubic feet. $\left(\frac{\pi}{4}=0\cdot7854.\right)$

4. Solve, by means of a graph,
$$3x+y=7,$$
$$x-3y=19.$$

5. A car sold for £435 yields a profit of 28 p.c. on the *selling pric*
Find the cost price.

6. If $a=4$, $b=5$, $c=6$, find the value of
$$\sqrt{a^2b+1}+\frac{2c}{a}-\frac{bc}{3}.$$

E

1. The manufacturer sells to the retailer at a price one-thir
above the price of manufacture, and the retailer charges the publ
one-third more than the article cost him. If £3. 16s. 0d. is charge
to the public, what was the cost of manufacture?　　(*R.S.A*

2. Factorise　　(*a*) $21(1-x^2)+40x$;
　　(*b*) $36+7x^2-4x^4$.

3. Solve　　$\frac{x}{2}-y=\frac{7}{2}$,
$$\frac{x}{25}+\frac{y}{2}=-\frac{3}{10}.$$

4. Find the fraction which becomes $\frac{3}{7}$ when 10 is added to i
numerator, and is equal to $\frac{1}{3}$ when 4 is taken from its denominatc

5. A ring is fixed 8 ft. above a horizontal floor. A and B a
points on the floor 6 ft. and 15 ft. respectively from the point whe
the ring would strike if it fell straight down. Calculate the length
the stretched string which passes through the ring and has its ends
A and B.

6. Given $6a=1$, $9b+1=0$, $2c=1$, find the value of
$$8a^3+27b^3+c^3-18abc.$$

F

1. A timber merchant has 200 pieces of boarding with an average length of 17 ft. 3 in. Of these 35 have an average length of 22 ft. while 125 pieces have an average length of 20 ft. What is the average length of the remaining pieces?

2. Solve
$$\frac{x-y}{2} - \frac{3x - \frac{1}{2}y}{5} = \frac{1}{2},$$

$$\frac{x}{3} - \frac{y}{2} = 2.$$

3. A sheet of copper 2 ft. 3 in. long, 1 ft. 3 in. broad and $\frac{1}{16}$ in. thick weighs 8 lb. Find to the nearest pound the weight of a cubic foot of copper.   (*R.S.A.*)

4. A 16-pint oil can costs 7*s.* 3*d.*, a 10-pint one costs 5*s.* 8*d.*, and one holding 2 pints costs 2*s.* 4*d.* Find by a graph the price of a 4-pint and of a gallon oil can.

5. Factorise   (*a*) $(a+b+c)^2 - 4\,(b-c)^2$;

(*b*) $2x^2 - xy - 6y^2$.

6. The metal in a solid sphere 7 in. diameter is melted down and cast into a solid cylinder 6 in. long. Find the diameter of the cylinder, allowing 5 p.c. for waste. ($\pi = 3\cdot14$.)

G

1. An alloy is made by mixing a lb. of metal worth x shillings per lb. with b lb. of another metal worth y shillings per lb. What is the value of the alloy in shillings per lb.?

2. Factorise   (*a*) $2x^2 + 5x - 12$;

(*b*) $x^2 + y^2 - z^2 + 2xy$.

3. Solve graphically
$$4x - y = 11,$$

$$\frac{5x}{2} + 2y = -1.$$

4. Find the volume of metal in a cast iron pipe, external diameter being 22 in., internal diameter being 20 in. and length 50 in. Give the answer in cubic inches. ($\pi = 3\frac{1}{7}$.)

5. $K = \dfrac{32Wlr^2}{\pi d^4}$. Change the subject to W, and find its value whe
$l = 80$, $d = 0.25$, $\pi = 3.142$, $2r = 2.55$, $K = 3.391 \times 10^7$.

6. A square steel plate is pierced by a square hole in the centr
of the plate, leaving a 4 in. margin. If the area of the hole is on
quarter that of the original plate, find the length of the side of th
plate.

H

1. If a case-hardening compound is bought at 41s. 8d. a cwt. an
sold at $5\frac{1}{2}d$. a pound, and there is a waste of 2 lb. per cwt. in handling
what percentage profit is made?

2. If $C = \dfrac{E - Cr}{R}$, change the subject to r, and find its value whe
$C = 1.5$, $E = 2.4$, $R = 1.15$.

3. Solve   (i) $17x - 19y = 15$,
                        $12x + 13y = 37$;

                    (ii) $7x^2 - 5x - 4.6 = 0$.

4. Evaluate by logs
$$\sqrt[3]{2.354 \times 1.607}.$$

5. Illustrate with diagrams the meaning of $\sin A$, $\cos A$, $\tan$
when A is an acute angle. If $\sin A = 0.6$, find the value of $\tan A$.

6. A passage in a building is 5 ft. wide and has a right-angle
turn. A man stands in the passage close to the outer wall and 15 f
from the corner of the building. How far can he see round th
turning measured from the outer corner along the outer wall of th
building?

If he walks along the passage towards the outer corner keepin
close to the outer wall, how far will he be from the corner when th
farthest point he can see along the other outer wall is the sam
distance from the corner as he is?

I

1. A tradesman's takings for a week are £75. What is his ne
profit, if he marks his goods 20 p.c. above the cost price and pays fo
rent 10 p.c. of his takings?

2. The following formula is used in an experiment in magnetism,

$$MH = \frac{4\pi^2 I}{T^2}.$$

Change the subject to T.

3. Find two values of t which will satisfy the equation

$$S = ut + \tfrac{1}{2}gt^2,$$

here $S = 336$, $u = 160$, $g = 32$.

4. Evaluate by logs $\quad \sqrt{\left(\dfrac{71\cdot6 \times 19\cdot29}{282\cdot1}\right)}$.

5. A shop has a frontage of 34 ft. An awning in front of the shop 34 ft. long and 8 ft. wide and is sloped at an angle of 28° to the orizontal. What area of pavement is protected from the rain falling ertically? What is the percentage decrease in this area if the angle increased to 36°? Illustrate your working by means of neat ketches.

6. (a) On a given straight line 2 in. long describe a segment of a ircle to contain an angle of 50°.

(b) Construct a triangle having its vertical angle 50°, a base ngle 60°, base 2 in.

J

1. A certain kind of brass is made up of 80 parts of copper, $12\tfrac{1}{2}$ arts of tin, and $2\tfrac{1}{2}$ parts of zinc. The price of copper is £80 per ton, f zinc £32 per ton, and of tin £112 per ton. Find to the nearest enny the cost of making a casting weighing 304 lb.

2. In each of the following express x in terms of y

(a) $y = mx + c$;

(b) $y = \dfrac{2 - 3x}{x + 3}$;

(c) $y = \dfrac{-1 \pm \sqrt{3x - 2}}{3}$,

3. Evaluate by logs

$$\pi \left\{ \frac{(3\tfrac{1}{2})^2 - (3\tfrac{1}{4})^2}{144} \right\}.$$

4. The diameters at the ends of a conical frustum are 5 in. an 3 in. respectively. Find the height of the frustum if its volume $187 cu. in. ($\pi = 3\frac{1}{7}$.)

5. Find the equation of the straight line passing through th points (2, 3) and (4, 5).

6. The elevation of a tower, taken from a point in the sam horizontal plane as its base is 60°. A flagstaff 15 ft. high on the to of the tower subtends an angle of 5° at the same station. Find th height of the tower.

K

1. A woman's pay is $\frac{3}{5}$ths of a man's, but the work done by woman is only $\frac{5}{8}$ths of a man's. If it costs £5 to pay for a man work, how much will it cost to pay a woman for doing the sam amount of work?

2. Draw on the same axes, and with same scale the graphs $y = \dfrac{30}{x^2}$ and $y = \dfrac{x}{3}$ from $x = 2$ to $x = 6$. What is the value of x at the intersection?

3. Evaluate by logs, given $\pi = 3 \cdot 14$, $g = 981$, $l = 18 \cdot 2$,

$$\frac{g}{4\pi^2 l}.$$

4. A bar of metal hexagonal in section is turned in a lathe, so a to obtain a cylindrical bar with the greatest possible cross-section By what percentage is the weight of the rod reduced?

5. The following formula is used to calculate the tonnage (T ton of a vessel of length (L ft.) and breadth (B ft.),

$$T = \tfrac{1}{188} (L - \tfrac{3}{5}B) B^2.$$

Express L in terms of T and B.

6. About a regular pentagon side $1\frac{1}{2}$ in., draw another whose side are 2 in. long and parallel to those of the smaller figure.

L

1. A mixture is made up of two kinds of metal costing respectivel 1s. 8d. and 1s. 2d. per lb. If the mixture contains 70 p.c. of th dearer kind, what p.c. profit is made by selling the mixture at 1s. 9 per lb.?

2. Evaluate by logs   (a) $\sqrt{\dfrac{0 \cdot 6165}{0 \cdot 47}}$;

(b) $\sqrt{(50)^2 + (83 \cdot 2)^2}$.

3. Divide a straight line into two parts so that the square on the larger part is equal to 4 times the product of the two parts. The line 15 in. long.

4. The longest and shortest sides of a right-angled triangle are 0 in. and 12 in. respectively. Find, both by scale drawing and calculation, the other side and the size of the smallest angle.

5. The cost of printing 200 copies is £3. 15s. 0d. and of printing 00 is £4. 7s. 6d. Find by a graph the cost of printing 450 copies.
(*C. and G.*)

6. Using your protractor, and without reference to your tables, nd the value of $3 \cos 20° + 4 \sin 20°$.

M

1. Given $t = 33$, and $a = \frac{1}{273}$, evaluate
$$\frac{760 \, (1 + at)}{\sqrt[3]{0 \cdot 04105}}.$$

2. A closed rectangular box whose external dimensions are 16 in. y 9 in. by 5 in., is made of wood $\frac{3}{8}$ in. thick and weighing 80 lb. per . ft. It contains 50 solid steel spheres of diameter 1 in. packed in wdust. The weight of the whole is 33·8 lb. Given that a cubic foot f steel weighs 500 lb., find the weight of a cubic foot of sawdust.
(*R.S.A.*)

3. Solve $\dfrac{x+2}{x-2} - \dfrac{x-2}{x+2} = \dfrac{24}{5}$.   (*R.A.F.*)

4. Draw the graph of $y = x^3 - 12x + 9$ from $x = +3$ to $x = -3$. For hat values of x is the gradient 0, and what is the gradient when) $x = 1$, (b) $x = -1$?

5. In the following change the subject from S to D:
$$S = \sqrt[3]{\dfrac{CPD^2}{3f}}.$$

6. 40 litres of a liquid costing 3s. 6d. per litre are mixed wit
110 litres of another liquid costing 2s. 0d. per litre. A contraction i
volume takes place on mixing equal to 4 p.c. of the volumes mixed
Find the cost of the mixture per litre.

N

1. A boiler and plate is 8 ft. 6 in. in diameter, and has one hole fo
a flue tube 3 ft. 6 in. in diameter. The thickness of the plate is $\frac{7}{8}$ i
Find its weight at 0·28 lb. per cu. in.

2. Solve

$$(a) \ \ x(6-x)=7;$$

$$(b) \ \ x+\frac{1}{y}=7\tfrac{2}{3},$$

$$2x+\frac{3}{y}=17\tfrac{5}{8}. \qquad \qquad (C. \ and \ G$$

3. If $bQ-\tfrac{1}{2}bP=a\left(P-\dfrac{Q}{2}\right)$, express P in terms of the othe
quantities, and find its value when $b=0\cdot375$, $a=1\cdot125$, $Q=0\cdot07$.

4. Evaluate by logs $D=\dfrac{5WL^3}{384EI}$, given that $W=4$, $L=200$
$E=12{,}500$, $I=18\cdot2$. $\qquad \qquad (N.C$

5. A builder employing 30 men, finds that he makes no profi
If he employs 60 men his profits are £20 per week. What are h
profits when he employs 40 men?

6. The angle of elevation of an unfinished chimney is 45°, th
point of observation being 130 ft. from its base. How much highe
must the chimney be built, so that the angle of elevation may be 56°

O

1. A square is inscribed in a circle whose circumference is $5\tfrac{1}{2}$ f
Find the length of the side of the square. $(\pi=3\tfrac{1}{7}.)$

2. If $P=\dfrac{17400t^2}{d\sqrt{L}}$, find the value of P when $t=\tfrac{1}{2}$, $L=17$, $d=35$.

$\qquad \qquad (N.C$

3. A box is 9·6 in. long, 7·2 in. wide and 5 in. deep. What is th
length of the longest piece of straight thin wire it will hold?

4. Simplify $\dfrac{x^2-8x+15}{x^2-7x+10} \div \dfrac{x^2+x-12}{x^2+2x-8}$.

5. Portable cast iron forges are listed at the following prices:

Area of hearth plate sq. ft.	7·2	4·3	2·2
Price in £	12·3	9·3	4·8

Find the price of a forge whose plate area is 4 sq. ft.　　(B.E.)

6. An object is dropped from an aeroplane 8000 ft. high. If it makes an angle of 3° with the vertical in its descent, how far from the spot vertically below the plane will it strike the ground?

P

1. Solve the following equations

(a) $3(2x-1)^2 + 7(2x-1) = 6$;

(b) $\dfrac{5}{7x} + \dfrac{1}{y} = 1$,

$\dfrac{1}{y} - \dfrac{11}{3x} = -14\tfrac{1}{3}$.

2. Experiments made with a *Weston differential pulley block* gave the following results, W being the load and P the effort, both in lb. Plot these results and find the relationship between P and W.

W	28	56	84	112	140	168	196
P	$7\tfrac{1}{4}$	$12\tfrac{1}{4}$	16	21	$25\tfrac{1}{2}$	$30\tfrac{1}{4}$	$34\tfrac{1}{4}$

3. A bolt consists of a square head and a cylindrical shank. The head is 2 in. square and 1 in. thick, and it is necessary that the weight of the shank shall be twice that of the head. Find

(a) the length of shank if its diameter is $1\tfrac{1}{2}$ in.;

(b) the diameter of the shank if its length is $4\tfrac{1}{2}$ in.

4. The jib of a crane is inclined at 57° to the horizontal; the post 12 ft. high and the tie-rod makes 35° with the horizontal and is 5 ft. long. Find (a) the length of the jib to the nearest inch, and (b) its height above the ground.

5. Given that $t = 33$, and $a = \frac{1}{273}$, evaluate

$$\frac{760\,(1 + at)}{\sqrt[3]{0 \cdot 04105}}.$$

6. OA and OB are two straight lines at right angles. A point A is 13 in. from O and its distance from OB is 2 in. more than twice its distance from OA. Find its distance from OA and OB.

Q

1. A cylindrical pipe $2\frac{1}{2}$ ft. in diameter discharges water at the rate of 3300 gal. per minute. Find the speed of water in the pipe (1) in feet per second, (2) in miles per hour. Take 1 cu. ft. $= 6\frac{1}{4}$ gal. and $\pi = 3\frac{1}{7}$.

2. Construct a regular pentagon having a diagonal of $2\frac{1}{2}$ in.

3. Solve by a graph

$$x + 2y = 2,$$
$$3y - 2x = 17.$$

4. Given $d = 50$, $\theta = 9°$, $l = 16 \cdot 4$, $T = 11 \cdot 5$, $I = 1750$, $\pi = 3 \cdot 142$,

$$\text{(i)}\quad \frac{M}{H} = \frac{\left\{ d^2 - \left(\frac{l}{2}\right)^2 \right\}^2 \tan\theta}{2d};$$

$$\text{(ii)}\quad MH = \frac{4\pi^2 I}{T^2}.$$

Find values of M and H.

5. A man buys a certain number of articles at $6d$. a dozen and an equal number at $9d$. a score. He sells the lot at $9d$. a dozen and makes a profit of $5s$. $6d$. How many did he buy?

6. Evaluate $\dfrac{(5x - 2y)^2 - (4x - 3y)^2}{9x^2 - 14xy + 5y^2}.$

Take $x = 2 \cdot 29$, $y = 1 \cdot 29$.

R

1. A pendulum is $3\frac{1}{2}$ ft. long and swings through an angle of 21. Another pendulum is 1 ft. 9 in. long and swings through an angle of 42°. Compare the areas covered by the two pendulums.

2. The average contents of 5 oil drums is $23\frac{1}{2}$ gal. Another drum is added which raises the average to 28 gal. per drum. How many gallons were in the sixth drum?

3. Solve   (a) $(x-1)(x-2)=(x-2)(x-3)$;

                   (b) $x(23-3x)=14$.

4. Draw a graph of the equation $y=x^2-x-6$ to show the values of y for values of x from $x=-1$ to $x=+3$. Find from your graph the value of x which gives y a minimum value.

5. A earns £5. 10s. 0d., while B earns 5 guineas, but A earns £3. 17s. 0d. in one day less than the number of days which B takes to earn £3. 17s. 0d. What are the daily wages of A and B?

6. The co-ordinates of a point are $(-5, 8)$. At what angle does a line, joining this point to the origin, cut the OX axis?

S

1. On the same axes draw the graphs of the following equations: (a) $y=2x^2-9x+30$, (b) $y=14+9x-2x^2$, between $x=0$, and $x=4$, taking an inch as unit for x and 0·2 in. as unit for y. From your graph, find the values of x which will make the two equations equal to one another.

2. $D=\dfrac{144}{134+r}$. Express r in terms of D and find its value when $D=0\cdot8$.   (N.C.)

3. In a triangle ABC, C being the right angle, AC is 12·2 in. and CB is 17·5 in. Find size of angle at B.   (B.E.)

4. A cast iron ball of 15 in. diameter is melted down and cast into conical mould of 20 in. diameter. Find height of cone.

5. Evaluate by logs

$$\sqrt[3]{\frac{812\cdot5 \times 3\cdot001}{78\cdot65^{\sqrt{3}}}}.$$

6. A car is made for £250. The manufacturer sells to an agent at a profit of 4 p.c., and the agent makes a profit of 15 p.c. on selling to the purchaser. What does the purchaser pay for it and how much per cent. is this above the original cost?

T

1. A number consists of two digits whose sum is 12. If one-thir of the number is increased by 1, the result is 4 times the tens digi Find the number.

2. A well is 12 ft. deep and 6 ft. in diameter. After pumping opera tions the water takes 30 min. to rise from the 9 ft. mark to the 11 f mark. Find its yield in gallons per minute. 1 cu. ft. = 6·23 gal.

3. A sum of money is invested partly at 4 p.c. and partly at $4\frac{1}{2}$ p. The total income from the two sources is £38. 2s. 0d. If the tota amount invested is £900, how much is invested under each head?

(G. and G

4. A line passing through the origin makes an angle of 33° 40 with the OX axis. If an x co-ordinate is 12, find the y co-ordinate the same point.

5. If $n^2r + 1 = NR$, express n in terms of r, N, R, and find its valu when $r = 6\frac{1}{4}$, $N = 24\frac{2}{3}$, $R = 1\frac{1}{2}$.

6. Draw the following graph, and find the equation connectin x and y.

x	4	8	12	20	24	48
y	6	9	12	18	21	39

(U.E.I

U

1. $F = \dfrac{WV^2}{gr}$. Express V in terms of the other quantities and fin its value when $W = 120$, $g = 32$, $F = 5$, $r = 48$.

2. Fig. 165 shows in *top row*: (a) Road Sign, (b) Angle Bricl (c) King Rod Roof Truss. In *bottom row*: (a) Hydrant Plate, (b) Hexa gon, (c) Plan of a Concrete Flat to a Bay Window.

Find the missing dimensions in each case.

3. If $\log a^{\frac{2}{3}} = 0·5126$, find a.

4. The angular points of a triangle are A (5, 5), B (18, 18), C (3, 15 Plot on squared paper and determine the equations of the side AB, BC, CA.

5. For what values of x will the following expression be equal to zero:
$$\tfrac{1}{4}\{x^2\,(x-3)^2-8x\,(x-3)-20\}.$$

6. A room is 36 ft. long, 15 ft. broad and 12 ft. high. What length of cord will reach from the centre of the floor to a corner of the ceiling? Answer correct to nearest foot.

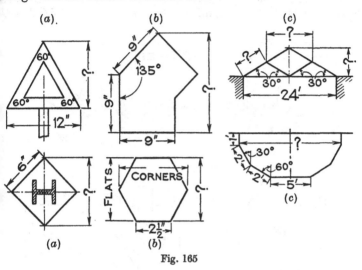

Fig. 165

V

1. A cast-iron surface plate 40 in. by 30 in. is being planed with a tool having a $\tfrac{3}{8}$ in. feed. How far does the tool travel in its *cutting stroke*?

2. A square thread of 2 in. pitch is being cut in a bar of 6 in. diameter. Find the volume of metal removed per pitch. See Fig. 166 (a).

3. Factorise
(i) r^2+r-12;
(ii) $D^2-256.$

4. In Fig. 166 (b) O is the centre of a circle, TN is a tangent. If the angle STN is $40°$, and BT and AC are perpendicular to each other, calculate the size of the angles CAB, ABT, and BTO.

5. (a) If sin $A = \frac{3}{5}$, find by calculation the values of cos A and tan A.

(b) Sand is piled against a wall so as to cover a strip of ground extending to 4 ft. in front of the wall. If the sand can rest with its surface inclined at 27° to the horizontal, find the volume of sand which can be piled in this way against every yard of the wall. Give the answer in cubic feet to 3 significant figures.   (N.C.T.E.C.)

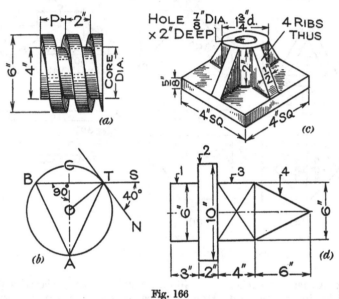

Fig. 166

6. (a) Find the weight of the footstep casting shown in Fig. 166 (c) given that cast-iron weighs 0·26 lb. per cu. in.

(b) Find the weight of the machined shaft shown in Fig. 166 (d) given that mild steel weighs 0·27 lb. per cu. in. Parts 1 and 2 are cylinders, part 3 is a square prism, part 4 is a cone.

TABLES AND CONSTANTS

Length
12 inches = 1 foot.
3 feet = 1 yard.
5½ yards = 1 rod, pole, or perch.
22 yards = 100 links = 1 chain.
40 poles = 1 furlong.
8 furlongs = 1760 yards
= 80 chains
= 1 mile.

Weight
16 oz. = 1 lb.
14 lb. = 1 stone.
28 lb. = 1 quarter (cwt.).
112 lb. = 1 cwt.
20 cwt. = 1 ton.

Square
144 sq. in. = 1 sq. ft.
9 sq. ft. = 1 sq. yd.
10 sq. chains = 1 acre.
4840 sq. yd. = 1 sq. mile
= 640 acres.

Cubic
1728 cu. in. = 1 cu. ft.
27 cu. ft. = 1 cu. yd.

Liquid and Dry Measures
4 gills = 1 pint.
2 pints = 1 quart.
4 quarts = 1 gallon.
2 gallons = 1 peck.
4 pecks = 1 bushel.

Metric Tables

Length
10 millimetres (mm.) = 1 centimetre (cm.).
10 centimetres = 1 decimetre (dcm.).
10 decimetres = **1 metre** (m.).
10 metres = 1 decametre.
10 decametres = 1 hectometre.
10 hectometres = **1 kilometre** (km.).

Measures of Area or Square Measure
100 sq. millimetres = 1 sq. centimetre (sq. cm.).
100 sq. centimetres = 1 sq. decimetre (sq. dcm.).
100 sq. decimetres = 1 sq. metre (sq. m.).
1 hectare = 100 ares = 10,000 sq. m.

Cubic Measure
1000 cubic millimetres = 1 cubic centimetre (1 c.c.).
1000 cubic centimetres = 1 cubic decimetre = **1 litre**.
10 milligrams = 1 centigram.
10 centigrams = 1 decigram.
10 decigrams = **1 gram** (gm.).
1000 grams = 1 kilogram (kilog.).
A gram is the weight of 1 c.c. of water at 4° C.

Conversion Table

1 centimetre = 0·01 m. = 0·394 in.

1 decimetre = 3·94 in. (3·937 more accurately).

1 metre = 39·37 in. = 3·281 ft. = 1·094 yd.

1 kilometre = 0·621 mile.

1 inch = 2·54 cm.

1 yard = 0·9144 metre (m.).

1 mile = 1609·3 m. or 1·6093 km.

1 sq. cm. = 0·155 sq. in.

1 sq. metre = 10·764 sq. ft. = 1·196 sq. yd.

1 sq. inch = 6·451 sq. cm.

1 sq. yard = 0·836 sq. metre.

1 litre = 1·76 pints = 0·22 gallon.

1 gallon = 0·1604 cu. ft. = 4·546 litres
= 10 lb. of water at 62° F.

$$\pi = 3\cdot14159 + \text{ or } 3\tfrac{1}{7}; \qquad \pi^2 = 9\cdot87.$$

$$\frac{1}{\pi} = 0\cdot3183; \qquad\qquad \frac{1}{\pi^2} = 0\cdot1013.$$

$$\sqrt{\pi} = 1\cdot7725; \qquad\qquad \frac{1}{\sqrt{\pi}} = 0\cdot564.$$

		Perimeter	Area
	Rectangle	$2\,(l+b)$	$l \times b$
	Square	$4 \times a$	a^2 or $\dfrac{d^2}{2}$
	Parallelo-gram	—	$a \times h$
	Triangle	—	$\dfrac{b \times h}{2}$
	Trapezium	—	$\tfrac{1}{2}\,(a+b) \times c$
	Triangle	$a+b+c$	If $s = \tfrac{1}{2}\,(a+b+c)$, Area $= \sqrt{s(s-a)(s-b)(s-c)}$
	Hexagon	$6s$ or $3 \cdot 46\,f$	$2 \cdot 598\,s^2$ or $0 \cdot 866\,f^2$
	Octagon	$8s$ or $3 \cdot 32\,f$	$4 \cdot 83\,s^2$ or $0 \cdot 829\,f^2$

		Perimeter	Area
	Circle	$2\pi r$ or πd	πr^2 or $0 \cdot 7854\, d^2$
	Hollow circle (Annulus)	—	$\pi\,(R+r)\,(R-r)$ $\dfrac{\pi}{4}\,(D+d)\,(D-d)$
	Hollow circle (Eccentric)	—	$\pi\,(R+r)\,(R-r)$ $\dfrac{\pi}{4}\,(D+d)\,(D-d)$
	Sector of circle	—	$\pi r^2 \times \dfrac{\theta}{360}$ or $\dfrac{\text{Arc} \times \text{radius}}{2}$
	Sector of hollow circle	—	$\pi\,(R^2 - r^2) \times \dfrac{\theta}{360}$
	Fillet	—	$0 \cdot 2146\, r^2$
	Radius	—	$0 \cdot 7854\, r^2$
	Segment of circle	—	Area of sector − area of triangle

		Surface area	Volume
	Any prism	Perimeter of base × vertical height + areas of ends	Area of base × vertical height
	Rectangular prism	$2\,(lb + lh + bh)$	$l \times b \times h$
	Cube	$(\text{Edge})^2 \times 6$	$(\text{Edge})^3$
	Square prism	$2\,(hb) + 4\,(bl)$	$b \times h \times l$
	Triangular prism	—	$\dfrac{b \times h}{2} \times l$
	Hexagonal prism	Lateral surface $= 6s \times h$ or $3{\cdot}46\,fh$	$2{\cdot}598\,s^2 \times h$ or $0{\cdot}866\,f^2 \times h$
	Octagonal prism	Lateral surface $= 8s \times h$ or $3{\cdot}32\,fh$	$4{\cdot}83\,s^2 \times h$ or $0{\cdot}829\,f^2 \times h$
	Cylinder	Lateral surface $= 2\pi rh$ Two ends $= 2\pi r^2$ Whole $= 2\pi r\,(r + h)$	$\pi r^2 h$ or $\pi d^2 h \div 4$
	Hollow cylinder	Outer lateral surface $= 2\pi Rh$ Inner lateral surface $= 2\pi rh$	$\pi\,(R^2 - r^2)\,h$ —

		Surface area	Volume
	Sphere	$4\pi R^2$	$\frac{4}{3}\pi R^3$ or $\frac{\pi}{6}D^3$ or $0\cdot5236\,D^3$
	Square pyramid	Lateral surface $= 2Sl$	$\frac{S^2H}{3}$
	Cone	Lateral surface $= \pi Rl$	$\frac{\pi R^2H}{3}$
	Frustum of square pyramid	Lateral surface $= 2l\,(S+s)$	$\frac{H}{3}\,(S^2+S^2+Ss)$
	Frustum of cone	Lateral surface $= \pi l\,(R+r)$	$\frac{\pi H}{3}\,(R^2+r^2+Rr)$
	Anchor ring	$4\pi^2 Rr$	$2\pi^2 Rr^2$

A Useful Table—Regular Polygons

Let $S =$ length of side of polygon.

Rule. Multiply S^2 by the decimal corresponding to the number of des in the polygon. The result is the area.

Number of sides	Multiply S^2 by
5	1·720
6	2·598
7	3·634
8	4·828
9	6·182
10	7·694
11	9·366
12	11·196

	0	1	2	3	4	5	6	7	8	9	Differences								
											1	2	3	4	5	6	7	8	9
10	·0000	0043	0086	0128	0170	0212	0253	0294	0334	0374	4	8	12	17	21	25	29	33	37
11	·0414	0453	0492	0531	0569	0607	0645	0682	0719	0755	4	8	11	15	19	23	26	30	34
12	·0792	0828	0864	0899	0934	0969	1004	1038	1072	1106	3	7	10	14	17	21	24	28	31
13	·1139	1173	1206	1239	1271	1303	1335	1367	1399	1430	3	6	10	13	16	19	23	26	29
14	·1461	1492	1523	1553	1584	1614	1644	1673	1703	1732	3	6	9	12	15	18	21	24	27
15	·1761	1790	1818	1847	1875	1903	1931	1959	1987	2014	3	6	8	11	14	17	20	22	25
16	·2041	2068	2095	2122	2148	2175	2201	2227	2253	2279	3	5	8	11	13	16	18	21	24
17	·2304	2330	2355	2380	2405	2430	2455	2480	2504	2529	2	5	7	10	12	15	17	20	22
18	·2553	2577	2601	2625	2648	2672	2695	2718	2742	2765	2	5	7	9	12	14	16	19	21
19	·2788	2810	2833	2856	2878	2900	2923	2945	2967	2989	2	4	7	9	11	13	16	18	20
20	·3010	3032	3054	3075	3096	3118	3139	3160	3181	3201	2	4	6	8	11	13	15	17	19
21	·3222	3243	3263	3284	3304	3324	3345	3365	3385	3404	2	4	6	8	10	12	14	16	18
22	·3424	3444	3464	3483	3502	3522	3541	3560	3579	3598	2	4	6	8	10	12	14	15	17
23	·3617	3636	3655	3674	3692	3711	3729	3747	3766	3784	2	4	6	7	9	11	13	15	17
24	·3802	3820	3838	3856	3874	3892	3909	3927	3945	3962	2	4	5	7	9	11	12	14	16
25	·3979	3997	4014	4031	4048	4065	4082	4099	4116	4133	2	3	5	7	9	10	12	14	15
26	·4150	4166	4183	4200	4216	4232	4249	4265	4281	4298	2	3	5	7	8	10	11	13	15
27	·4314	4330	4346	4362	4378	4393	4409	4425	4440	4456	2	3	5	6	8	9	11	13	14
28	·4472	4487	4502	4518	4533	4548	4564	4579	4594	4609	2	3	5	6	8	9	11	12	14
29	·4624	4639	4654	4669	4683	4698	4713	4728	4742	4757	1	3	4	6	7	9	10	12	13
30	·4771	4786	4800	4814	4829	4843	4857	4871	4886	4900	1	3	4	6	7	9	10	11	13
31	·4914	4928	4942	4955	4969	4983	4997	5011	5024	5038	1	3	4	6	7	8	10	11	12
32	·5051	5065	5079	5092	5105	5119	5132	5145	5159	5172	1	3	4	5	7	8	9	11	12
33	·5185	5198	5211	5224	5237	5250	5263	5276	5289	5302	1	3	4	5	6	8	9	10	12
34	·5315	5328	5340	5353	5366	5378	5391	5403	5416	5428	1	3	4	5	6	8	9	10	11
35	·5441	5453	5465	5478	5490	5502	5514	5527	5539	5551	1	2	4	5	6	7	9	10	11
36	·5563	5575	5587	5599	5611	5623	5635	5647	5658	5670	1	2	4	5	6	7	8	10	11
37	·5682	5694	5705	5717	5729	5740	5752	5763	5775	5786	1	2	3	5	6	7	8	9	10
38	·5798	5809	5821	5832	5843	5855	5866	5877	5888	5899	1	2	3	5	6	7	8	9	10
39	·5911	5922	5933	5944	5955	5966	5977	5988	5999	6010	1	2	3	4	5	7	8	9	10
40	·6021	6031	6042	6053	6064	6075	6085	6096	6107	6117	1	2	3	4	5	6	8	9	10
41	·6128	6138	6149	6160	6170	6180	6191	6201	6212	6222	1	2	3	4	5	6	7	8	9
42	·6232	6243	6253	6263	6274	6284	6294	6304	6314	6325	1	2	3	4	5	6	7	8	9
43	·6335	6345	6355	6365	6375	6385	6395	6405	6415	6425	1	2	3	4	5	6	7	8	9
44	·6435	6444	6454	6464	6474	6484	6493	6503	6513	6522	1	2	3	4	5	6	7	8	9
45	·6532	6542	6551	6561	6571	6580	6590	6599	6609	6618	1	2	3	4	5	6	7	8	9
46	·6628	6637	6646	6656	6665	6675	6684	6693	6702	6712	1	2	3	4	5	6	7	7	8
47	·6721	6730	6739	6749	6758	6767	6776	6785	6794	6803	1	2	3	4	5	5	6	7	8
48	·6812	6821	6830	6839	6848	6857	6866	6875	6884	6893	1	2	3	4	4	5	6	7	8
49	·6902	6911	6920	6928	6937	6946	6955	6964	6972	6981	1	2	3	4	4	5	6	7	8
50	·6990	6998	7007	7016	7024	7033	7042	7050	7059	7067	1	2	3	3	4	5	6	7	8
51	·7076	7084	7093	7101	7110	7118	7126	7135	7143	7152	1	2	3	3	4	5	6	7	8
52	·7160	7168	7177	7185	7193	7202	7210	7218	7226	7235	1	2	2	3	4	5	6	7	7
53	·7243	7251	7259	7267	7275	7284	7292	7300	7308	7316	1	2	2	3	4	5	6	6	7
54	·7324	7332	7340	7348	7356	7364	7372	7380	7388	7396	1	2	2	3	4	5	6	6	7

	0	1	2	3	4	5	6	7	8	9	1 2 3	4 5 6	7 8 9
55	·7404	7412	7419	7427	7435	7443	7451	7459	7466	7474	1 2 2	3 4 5	5 6 7
56	·7482	7490	7497	7505	7513	7520	7528	7536	7543	7551	1 2 2	3 4 5	5 6 7
57	·7559	7566	7574	7582	7589	7597	7604	7612	7619	7627	1 2 2	3 4 5	5 6 7
58	·7634	7642	7649	7657	7664	7672	7679	7686	7694	7701	1 1 2	3 4 4	5 6 7
59	·7709	7716	7723	7731	7738	7745	7752	7760	7767	7774	1 1 2	3 4 4	5 6 7
60	·7782	7789	7796	7803	7810	7818	7825	7832	7839	7846	1 1 2	3 4 4	5 6 6
61	·7853	7860	7868	7875	7882	7889	7896	7903	7910	7917	1 1 2	3 4 4	5 6 6
62	·7924	7931	7938	7945	7952	7959	7966	7973	7980	7987	1 1 2	3 4 4	5 6 6
63	·7993	8000	8007	8014	8021	8028	8035	8041	8048	8055	1 1 2	3 3 4	5 5 6
64	·8062	8069	8075	8082	8089	8096	8102	8109	8116	8122	1 1 2	3 3 4	5 5 6
65	·8129	8136	8142	8149	8156	8162	8169	8176	8182	8189	1 1 2	3 3 4	5 5 6
66	·8195	8202	8209	8215	8222	8228	8235	8241	8248	8254	1 1 2	3 3 4	5 5 6
67	·8261	8267	8274	8280	8287	8293	8299	8306	8312	8319	1 1 2	3 3 4	5 5 6
68	·8325	8331	8338	8344	8351	8357	8363	8370	8376	8382	1 1 2	3 3 4	4 5 6
69	·8388	8395	8401	8407	8414	8420	8426	8432	8439	8445	1 1 2	2 3 4	4 5 6
70	·8451	8457	8463	8470	8476	8482	8488	8494	8500	8506	1 1 2	2 3 4	4 5 6
71	·8513	8519	8525	8531	8537	8543	8549	8555	8561	8567	1 1 2	2 3 4	4 5 5
72	·8573	8579	8585	8591	8597	8603	8609	8615	8621	8627	1 1 2	2 3 4	4 5 5
73	·8633	8639	8645	8651	8657	8663	8669	8675	8681	8686	1 1 2	2 3 4	4 5 5
74	·8692	8698	8704	8710	8716	8722	8727	8733	8739	8745	1 1 2	2 3 4	4 5 5
75	·8751	8756	8762	8768	8774	8779	8785	8791	8797	8802	1 1 2	2 3 3	4 5 5
76	·8808	8814	8820	8825	8831	8837	8842	8848	8854	8859	1 1 2	2 3 3	4 5 5
77	·8865	8871	8876	8882	8887	8893	8899	8904	8910	8915	1 1 2	2 3 3	4 4 5
78	·8921	8927	8932	8938	8943	8949	8954	8960	8965	8971	1 1 2	2 3 3	4 4 5
79	·8976	8982	8987	8993	8998	9004	9009	9015	9020	9025	1 1 2	2 3 3	4 4 5
80	·9031	9036	9042	9047	9053	9058	9063	9069	9074	9079	1 1 2	2 3 3	4 4 5
81	·9085	9090	9096	9101	9106	9112	9117	9122	9128	9133	1 1 2	2 3 3	4 4 5
82	·9138	9143	9149	9154	9159	9165	9170	9175	9180	9186	1 1 2	2 3 3	4 4 5
83	·9191	9196	9201	9206	9212	9217	9222	·9227	9232	9238	1 1 2	2 3 3	4 4 5
84	·9243	9248	9253	9258	9263	9269	9274	9279	9284	9289	1 1 2	2 3 3	4 4 5
85	·9294	9299	9304	9309	9315	9320	9325	9330	9335	9340	1 1 2	2 3 3	4 4 5
86	·9345	9350	9355	9360	9365	9370	9375	9380	9385	9390	1 1 2	2 3 3	4 4 5
87	·9395	9400	9405	9410	9415	9420	9425	9430	9435	9440	0 1 1	2 2 3	3 4 4
88	·9445	9450	9455	9460	9465	9469	9474	9479	9484	9489	0 1 1	2 2 3	3 4 4
89	·9494	9499	9504	9509	9513	9518	9523	9528	9533	9538	0 1 1	2 2 3	3 4 4
90	·9542	9547	9552	9557	9562	9566	9571	9576	9581	9586	0 1 1	2 2 3	3 4 4
91	·9590	9595	9600	9605	9609	9614	9619	9624	9628	9633	0 1 1	2 2 3	3 4 4
92	·9638	9643	9647	9652	9657	9661	9666	9671	9675	9680	0 1 1	2 2 3	3 4 4
93	·9685	9689	9694	9699	9703	9708	9713	9717	9722	9727	0 1 1	2 2 3	3 4 4
94	·9731	9736	9741	9745	9750	9754	9759	9763	9768	9773	0 1 1	2 2 3	3 4 4
95	·9777	9782	9786	9791	9795	9800	9805	9809	9814	9818	0 1 1	2 2 3	3 4 4
96	·9823	9827	9832	9836	9841	9845	9850	9854	9859	9863	0 1 1	2 2 3	3 4 4
97	·9868	9872	9877	9881	9886	9890	9894	9899	9903	9908	0 1 1	2 2 3	3 4 4
98	·9912	9917	9921	9926	9930	9934	9939	9943	9948	9952	0 1 1	2 2 3	3 4 4
99	·9956	9961	9965	9969	9974	9978	9983	9987	9991	9996	0 1 1	2 2 3	3 3 4

Differences

	0	1	2	3	4	5	6	7	8	9	Differences								
											1	2	3	4	5	6	7	8	9
·00	1000	1002	1005	1007	1009	1012	1014	1016	1019	1021	0 0 1			1 1 1			2 2 2		
·01	1023	1026	1028	1030	1033	1035	1038	1040	1042	1045	0 0 1			1 1 1			2 2 2		
·02	1047	1050	1052	1054	1057	1059	1062	1064	1067	1069	0 0 1			1 1 1			2 2 2		
·03	1072	1074	1076	1079	1081	1084	1086	1089	1091	1094	0 0 1			1 1 1			2 2 2		
·04	1096	1099	1102	1104	1107	1109	1112	1114	1117	1119	0 1 1			1 1 2			2 2 2		
·05	1122	1125	1127	1130	1132	1135	1138	1140	1143	1146	0 1 1			1 1 2			2 2 2		
·06	1148	1151	1153	1156	1159	1161	1164	1167	1169	1172	0 1 1			1 1 2			2 2 2		
·07	1175	1178	1180	1183	1186	1189	1191	1194	1197	1199	0 1 1			1 1 2			2 2 2		
·08	1202	1205	1208	1211	1213	1216	1219	1222	1225	1227	0 1 1			1 1 2			2 2 3		
·09	1230	1233	1236	1239	1242	1245	1247	1250	1253	1256	0 1 1			1 1 2			2 2 3		
·10	1259	1262	1265	1268	1271	1274	1276	1279	1282	1285	0 1 1			1 1 2			2 2 3		
·11	1288	1291	1294	1297	1300	1303	1306	1309	1312	1315	0 1 1			1 2 2			2 2 3		
·12	1318	1321	1324	1327	1330	1334	1337	1340	1343	1346	0 1 1			1 2 2			2 2 3		
·13	1349	1352	1355	1358	1361	1365	1368	1371	1374	1377	0 1 1			1 2 2			2 3 3		
·14	1380	1384	1387	1390	1393	1396	1400	1403	1406	1409	0 1 1			1 2 2			2 3 3		
·15	1413	1416	1419	1422	1426	1429	1432	1435	1439	1442	0 1 1			1 2 2			2 3 3		
·16	1445	1449	1452	1455	1459	1462	1466	1469	1472	1476	0 1 1			1 2 2			2 3 3		
·17	1479	1483	1486	1489	1493	1496	1500	1503	1507	1510	0 1 1			1 2 2			2 3 3		
·18	1514	1517	1521	1524	1528	1531	1535	1538	1542	1545	0 1 1			1 2 2			2 3 3		
·19	1549	1552	1556	1560	1563	1567	1570	1574	1578	1581	0 1 1			1 2 2			3 3 3		
·20	1585	1589	1592	1596	1600	1603	1607	1611	1614	1618	0 1 1			1 2 2			3 3 3		
·21	1622	1626	1629	1633	1637	1641	1644	1648	1652	1656	0 1 1			2 2 2			3 3 3		
·22	1660	1663	1667	1671	1675	1679	1683	1687	1690	1694	0 1 1			2 2 2			3 3 3		
·23	1698	1702	1706	1710	1714	1718	1722	1726	1730	1734	0 1 1			2 2 2			3 3 4		
·24	1738	1742	1746	1750	1754	1758	1762	1766	1770	1774	0 1 1			2 2 2			3 3 4		
·25	1778	1782	1786	1791	1795	1799	1803	1807	1811	1816	0 1 1			2 2 2			3 3 4		
·26	1820	1824	1828	1832	1837	1841	1845	1849	1854	1858	0 1 1			2 2 3			3 3 4		
·27	1862	1866	1871	1875	1879	1884	1888	1892	1897	1901	0 1 1			2 2 3			3 3 4		
·28	1905	1910	1914	1919	1923	1928	1932	1936	1941	1945	0 1 1			2 2 3			3 4 4		
·29	1950	1954	1959	1963	1968	1972	1977	1982	1986	1991	0 1 1			2 2 3			3 4 4		
·30	1995	2000	2004	2009	2014	2018	2023	2028	2032	2037	0 1 1			2 2 3			3 4 4		
·31	2042	2046	2051	2056	2061	2065	2070	2075	2080	2084	0 1 1			2 2 3			3 4 4		
·32	2089	2094	2099	2104	2109	2113	2118	2123	2128	2133	0 1 1			2 2 3			3 4 4		
·33	2138	2143	2148	2153	2158	2163	2168	2173	2178	2183	0 1 1			2 2 3			3 4 4		
·34	2188	2193	2198	2203	2208	2213	2218	2223	2228	2234	1 1 2			2 3 3			4 4 5		
·35	2239	2244	2249	2254	2259	2265	2270	2275	2280	2286	1 1 2			2 3 3			4 4 5		
·36	2291	2296	2301	2307	2312	2317	2323	2328	2333	2339	1 1 2			2 3 3			4 4 5		
·37	2344	2350	2355	2360	2366	2371	2377	2382	2388	2393	1 1 2			2 3 3			4 4 5		
·38	2399	2404	2410	2415	2421	2427	2432	2438	2443	2449	1 1 2			2 3 3			4 4 5		
·39	2455	2460	2466	2472	2477	2483	2489	2495	2500	2506	1 1 2			2 3 3			4 5 5		
·40	2512	2518	2523	2529	2535	2541	2547	2553	2559	2564	1 1 2			2 3 4			4 5 5		
·41	2570	2576	2582	2588	2594	2600	2606	2612	2618	2624	1 1 2			2 3 4			4 5 5		
·42	2630	2636	2642	2649	2655	2661	2667	2673	2679	2685	1 1 2			2 3 4			4 5 6		
·43	2692	2698	2704	2710	2716	2723	2729	2735	2742	2748	1 1 2			3 3 4			4 5 6		
·44	2754	2761	2767	2773	2780	2786	2793	2799	2805	2812	1 1 2			3 3 4			4 5 6		
·45	2818	2825	2831	2838	2844	2851	2858	2864	2871	2877	1 1 2			3 3 4			5 5 6		
·46	2884	2891	2897	2904	2911	2917	2924	2931	2938	2944	1 1 2			3 3 4			5 5 6		
·47	2951	2958	2965	2972	2979	2985	2992	2999	3006	3013	1 1 2			3 3 4			5 5 6		
·48	3020	3027	3034	3041	3048	3055	3062	3069	3076	3083	1 1 2			3 4 4			5 6 6		
·49	3090	3097	3105	3112	3119	3126	3133	3141	3148	3155	1 1 2			3 4 4			5 6 6		

	0	1	2	3	4	5	6	7	8	9	Differences								
											1	2	3	4	5	6	7	8	9
50	3162	3170	3177	3184	3192	3199	3206	3214	3221	3228	1	1	2	3	4	4	5	6	7
51	3236	3243	3251	3258	3266	3273	3281	3289	3296	3304	1	2	2	3	4	5	5	6	7
52	3311	3319	3327	3334	3342	3350	3357	3365	3373	3381	1	2	2	3	4	5	5	6	7
53	3388	3396	3404	3412	3420	3428	3436	3443	3451	3459	1	2	2	3	4	5	6	6	7
54	3467	3475	3483	3491	3499	3508	3516	3524	3532	3540	1	2	2	3	4	5	6	6	7
55	3548	3556	3565	3573	3581	3589	3597	3606	3614	3622	1	2	2	3	4	5	6	7	7
56	3631	3639	3648	3656	3664	3673	3681	3690	3698	3707	1	2	3	3	4	5	6	7	8
57	3715	3724	3733	3741	3750	3758	3767	3776	3784	3793	1	2	3	3	4	5	6	7	8
58	3802	3811	3819	3828	3837	3846	3855	3864	3873	3882	1	2	3	4	4	5	6	7	8
59	3890	3899	3908	3917	3926	3936	3945	3954	3963	3972	1	2	3	4	5	5	6	7	8
60	3981	3990	3999	4009	4018	4027	4036	4046	4055	4064	1	2	3	4	5	6	6	7	8
61	4074	4083	4093	4102	4111	4121	4130	4140	4150	4159	1	2	3	4	5	6	7	8	9
62	4169	4178	4188	4198	4207	4217	4227	4236	4246	4256	1	2	3	4	5	6	7	8	9
63	4266	4276	4285	4295	4305	4315	4325	4335	4345	4355	1	2	3	4	5	6	7	8	9
64	4365	4375	4385	4395	4406	4416	4426	4436	4446	4457	1	2	3	4	5	6	7	8	9
65	4467	4477	4487	4498	4508	4519	4529	4539	4550	4560	1	2	3	4	5	6	7	8	9
66	4571	4581	4592	4603	4613	4624	4634	4645	4656	4667	1	2	3	4	5	6	7	8	10
67	4677	4688	4699	4710	4721	4732	4742	4753	4764	4775	1	2	3	4	5	7	8	9	10
68	4786	4797	4808	4819	4831	4842	4853	4864	4875	4887	1	2	3	4	6	7	8	9	10
69	4898	4909	4920	4932	4943	4955	4966	4977	4989	5000	1	2	3	5	6	7	8	9	10
70	5012	5023	5035	5047	5058	5070	5082	5093	5105	5117	1	2	4	5	6	7	8	9	11
71	5129	5140	5152	5164	5176	5188	5200	5212	5224	5236	1	2	4	5	6	7	8	10	11
72	5248	5260	5272	5284	5297	5309	5321	5333	5346	5358	1	2	4	5	6	7	9	10	11
73	5370	5383	5395	5408	5420	5433	5445	5458	5470	5483	1	3	4	5	6	8	9	10	11
74	5495	5508	5521	5534	5546	5559	5572	5585	5598	5610	1	3	4	5	6	8	9	10	12
75	5623	5636	5649	5662	5675	5689	5702	5715	5728	5741	1	3	4	5	7	8	9	10	12
76	5754	5768	5781	5794	5808	5821	5834	5848	5861	5875	1	3	4	5	7	8	9	11	12
77	5888	5902	5916	5929	5943	5957	5970	5984	5998	6012	1	3	4	5	7	8	10	11	12
78	6026	6039	6053	6067	6081	6095	6109	6124	6138	6152	1	3	4	6	7	8	10	11	13
79	6166	6180	6194	6209	6223	6237	6252	6266	6281	6295	1	3	4	6	7	9	10	11	13
80	6310	6324	6339	6353	6368	6383	6397	6412	6427	6442	1	3	4	6	7	9	10	12	13
81	6457	6471	6486	6501	6516	6531	6546	6561	6577	6592	2	3	5	6	8	9	11	12	14
82	6607	6622	6637	6653	6668	6683	6699	6714	6730	6745	2	3	5	6	8	9	11	12	14
83	6761	6776	6792	6808	6823	6839	6855	6871	6887	6902	2	3	5	6	8	9	11	13	14
84	6918	6934	6950	6966	6982	6998	7015	7031	7047	7063	2	3	5	6	8	10	11	13	15
85	7079	7096	7112	7129	7145	7161	7178	7194	7211	7228	2	3	5	7	8	10	12	13	15
86	7244	7261	7278	7295	7311	7328	7345	7362	7379	7396	2	3	5	7	8	10	12	13	15
87	7413	7430	7447	7464	7482	7499	7516	7534	7551	7568	2	3	5	7	9	10	12	14	16
88	7586	7603	7621	7638	7656	7674	7691	7709	7727	7745	2	4	5	7	9	11	12	14	16
89	7762	7780	7798	7816	7834	7852	7870	7889	7907	7925	2	4	5	7	9	11	13	14	16
90	7943	7962	7980	7998	8017	8035	8054	8072	8091	8110	2	4	6	7	9	11	13	15	17
91	8128	8147	8166	8185	8204	8222	8241	8260	8279	8299	2	4	6	8	9	11	13	15	17
92	8318	8337	8356	8375	8395	8414	8433	8453	8472	8492	2	4	6	8	10	12	14	15	17
93	8511	8531	8551	8570	8590	8610	8630	8650	8670	8690	2	4	6	8	10	12	14	16	18
94	8710	8730	8750	8770	8790	8810	8831	8851	8872	8892	2	4	6	8	10	12	14	16	18
95	8913	8933	8954	8974	8995	9016	9036	9057	9078	9099	2	4	6	8	10	12	14	17	19
96	9120	9141	9162	9183	9204	9226	9247	9268	9290	9311	2	4	6	8	11	13	15	17	19
97	9333	9354	9376	9397	9419	9441	9462	9484	9506	9528	2	4	7	9	11	13	15	17	20
98	9550	9572	9594	9616	9638	9661	9683	9705	9727	9750	2	4	7	9	11	13	16	18	20
99	9772	9795	9817	9840	9863	9886	9908	9931	9954	9977	2	5	7	9	11	14	16	18	20

	0′	6′	12′	18′	24′	30′	36′	42′	48′	54′	1′	2′	3′	4′	5′
0°	·0000	0017	0035	0052	0070	0087	0105	0122	0140	0157	3	6	9	12	1
1	·0175	0192	0209	0227	0244	0262	0279	0297	0314	0332	3	6	9	12	1
2	·0349	0366	0384	0401	0419	0436	0454	0471	0488	0506	3	6	9	12	1
3	·0523	0541	0558	0576	0593	0610	0628	0645	0663	0680	3	6	9	12	1
4	·0698	0715	0732	0750	0767	0785	0802	0819	0837	0854	3	6	9	12	14
5	·0872	0889	0906	0924	0941	0958	0976	0993	1011	1028	3	6	9	12	14
6	·1045	1063	1080	1097	1115	1132	1149	1167	1184	1201	3	6	9	12	14
7	·1219	1236	1253	1271	1288	1305	1323	1340	1357	1374	3	6	9	12	14
8	·1392	1409	1426	1444	1461	1478	1495	1513	1530	1547	3	6	9	12	14
9	·1564	1582	1599	1616	1633	1650	1668	1685	1702	1719	3	6	9	11	14
10	·1736	1754	1771	1788	1805	1822	1840	1857	1874	1891	3	6	9	11	14
11	·1908	1925	1942	1959	1977	1994	2011	2028	2045	2062	3	6	9	11	14
12	·2079	2096	2113	2130	2147	2164	2181	2198	2215	2233	3	6	9	11	14
13	·2250	2267	2284	2300	2317	2334	2351	2368	2385	2402	3	6	8	11	14
14	·2419	2436	2453	2470	2487	2504	2521	2538	2554	2571	3	6	8	11	14
15	·2588	2605	2622	2639	2656	2672	2689	2706	2723	2740	3	6	8	11	14
16	·2756	2773	2790	2807	2823	2840	2857	2874	2890	2907	3	6	8	11	14
17	·2924	2940	2957	2974	2990	3007	3024	3040	3057	3074	3	6	8	11	14
18	·3090	3107	3123	3140	3156	3173	3190	3206	3223	3239	3	6	8	11	14
19	·3256	3272	3289	3305	3322	3338	3355	3371	3387	3404	3	5	8	11	14
20	·3420	3437	3453	3469	3486	3502	3518	3535	3551	3567	3	5	8	11	14
21	·3584	3600	3616	3633	3649	3665	3681	3697	3714	3730	3	5	8	11	14
22	·3746	3762	3778	3795	3811	3827	3843	3859	3875	3891	3	5	8	11	13
23	·3907	3923	3939	3955	3971	3987	4003	4019	4035	4051	3	5	8	11	13
24	·4067	4083	4099	4115	4131	4147	4163	4179	4195	4210	3	5	8	11	13
25	·4226	4242	4258	4274	4289	4305	4321	4337	4352	4368	3	5	8	11	13
26	·4384	4399	4415	4431	4446	4462	4478	4493	4509	4524	3	5	8	10	13
27	·4540	4555	4571	4586	4602	4617	4633	4648	4664	4679	3	5	8	10	13
28	·4695	4710	4726	4741	4756	4772	4787	4802	4818	4833	3	5	8	10	13
29	·4848	4863	4879	4894	4909	4924	4939	4955	4970	4985	3	5	8	10	13
30	·5000	5015	5030	5045	5060	5075	5090	5105	5120	5135	3	5	8	10	13
31	·5150	5165	5180	5195	5210	5225	5240	5255	5270	5284	2	5	7	10	12
32	·5299	5314	5329	5344	5358	5373	5388	5402	5417	5432	2	5	7	10	12
33	·5446	5461	5476	5490	5505	5519	5534	5548	5563	5577	2	5	7	10	12
34	·5592	5606	5621	5635	5650	5664	5678	5693	5707	5721	2	5	7	10	12
35	·5736	5750	5764	5779	5793	5807	5821	5835	5850	5864	2	5	7	9	12
36	·5878	5892	5906	5920	5934	5948	5962	5976	5990	6004	2	5	7	9	12
37	·6018	6032	6046	6060	6074	6088	6101	6115	6129	6143	2	5	7	9	1
38	·6157	6170	6184	6198	6211	6225	6239	6252	6266	6280	2	5	7	9	11
39	·6293	6307	6320	6334	6347	6361	6374	6388	6401	6414	2	4	7	9	11
40	·6428	6441	6455	6468	6481	6494	6508	6521	6534	6547	2	4	7	9	11
41	·6561	6574	6587	6600	6613	6626	6639	6652	6665	6678	2	4	7	9	11
42	·6691	6704	6717	6730	6743	6756	6769	6782	6794	6807	2	4	6	9	11
43	·6820	6833	6845	6858	6871	6884	6896	6909	6921	6934	2	4	6	8	11
44	·6947	6959	6972	6984	6997	7009	7022	7034	7046	7059	2	4	6	8	10
	0′	6′	12′	18′	24′	30′	36′	42′	48′	54′	1′	2′	3′	4′	5′

	0'	6'	12'	18'	24'	30'	36'	42'	48'	54'	1'	2'	3'	4'	5'
45°	·7071	7083	7096	7108	7120	7133	7145	7157	7169	7181	2	4	6	8	10
46	·7193	7206	7218	7230	7242	7254	7266	7278	7290	7302	2	4	6	8	10
47	·7314	7325	7337	7349	7361	7373	7385	7396	7408	7420	2	4	6	8	10
48	·7431	7443	7455	7466	7478	7490	7501	7513	7524	7536	2	4	6	8	10
49	·7547	7559	7570	7581	7593	7604	7615	7627	7638	7649	2	4	6	8	9
50	·7660	7672	7683	7694	7705	7716	7727	7738	7749	7760	2	4	6	7	9
51	·7771	7782	7793	7804	7815	7826	7837	7848	7859	7869	2	4	5	7	9
52	·7880	7891	7902	7912	7923	7934	7944	7955	7965	7976	2	4	5	7	9
53	·7986	7997	8007	8018	8028	8039	8049	8059	8070	8080	2	3	5	7	9
54	·8090	8100	8111	8121	8131	8141	8151	8161	8171	8181	2	3	5	7	8
55	·8192	8202	8211	8221	8231	8241	8251	8261	8271	8281	2	3	5	7	8
56	·8290	8300	8310	8320	8329	8339	8348	8358	8368	8377	2	3	5	6	8
57	·8387	8396	8406	8415	8425	8434	8443	8453	8462	8471	2	3	5	6	8
58	·8480	8490	8499	8508	8517	8526	8536	8545	8554	8563	2	3	5	6	8
59	·8572	8581	8590	8599	8607	8616	8625	8634	8643	8652	1	3	4	6	7
60	·8660	8669	8678	8686	8695	8704	8712	8721	8729	8738	1	3	4	6	7
61	·8746	8755	8763	8771	8780	8788	8796	8805	8813	8821	1	3	4	6	7
62	·8829	8838	8846	8854	8862	8870	8878	8886	8894	8902	1	3	4	5	7
63	·8910	8918	8926	8934	8942	8949	8957	8965	8973	8980	1	3	4	5	6
64	·8988	8996	9003	9011	9018	9026	9033	9041	9048	9056	1	3	4	5	6
65	·9063	9070	9078	9085	9092	9100	9107	9114	9121	9128	1	2	4	5	6
66	·9135	9143	9150	9157	9164	9171	9178	9184	9191	9198	1	2	3	5	6
67	·9205	9212	9219	9225	9232	9239	9245	9252	9259	9265	1	2	3	4	6
68	·9272	9278	9285	9291	9298	9304	9311	9317	9323	9330	1	2	3	4	5
69	·9336	9342	9348	9354	9361	9367	9373	9379	9385	9391	1	2	3	4	5
70	·9397	9403	9409	9415	9421	9426	9432	9438	9444	9449	1	2	3	4	5
71	·9455	9461	9466	9472	9478	9483	9489	9494	9500	9505	1	2	3	4	5
72	·9511	9516	9521	9527	9532	9537	9542	9548	9553	9558	1	2	3	4	4
73	·9563	9568	9573	9578	9583	9588	9593	9598	9603	9608	1	2	2	3	4
74	·9613	9617	9622	9627	9632	9636	9641	9646	9650	9655	1	2	2	3	4
75	·9659	9664	9668	9673	9677	9681	9686	9690	9694	9699	1	1	2	3	4
76	·9703	9707	9711	9715	9720	9724	9728	9732	9736	9740	1	1	2	2	3
77	·9744	9748	9751	9755	9759	9763	9767	9770	9774	9778	1	1	2	3	3
78	·9781	9785	9789	9792	9796	9799	9803	9806	9810	9813	1	1	2	2	3
79	·9816	9820	9823	9826	9829	9833	9836	9839	9842	9845	1	1	2	2	3
80	·9848	9851	9854	9857	9860	9863	9866	9869	9871	9874	0	1	1	2	2
81	·9877	9880	9882	9885	9888	9890	9893	9895	9898	9900	0	1	1	2	2
82	·9903	9905	9907	9910	9912	9914	9917	9919	9921	9923	0	1	1	2	2
83	·9925	9928	9930	9932	9934	9936	9938	9940	9942	9943	0	1	1	1	2
84	·9945	9947	9949	9951	9952	9954	9956	9957	9959	9960	0	1	1	1	1
85	·9962	9963	9965	9966	9968	9969	9971	9972	9973	9974	0	0	1	1	1
86	·9976	9977	9978	9979	9980	9981	9982	9983	9984	9985	0	0	1	1	1
87	·9986	9987	9988	9989	9990	9990	9991	9992	9993	9993					
88	·9994	9995	9995	9996	9996	9997	9997	9997	9998	9998					
89	·9998	9999	9999	9999	9999	1·000	1·000	1·000	1·000	1·000					
	0'	6'	12'	18'	24'	30'	36'	42'	48'	54'	1'	2'	3'	4'	5'

	0'	6'	12'	18'	24'	30'	36'	42'	48'	54'	1'	2'	3'	4'	5'
0°	1·0000	1·0000	1·0000	1·0000	1·0000	1·0000	**9999**	**9999**	**9999**	**9999**					
1	·9998	9998	9998	9997	9997	9997	9996	9996	9995	9995					
2	·9994	9993	9993	9992	9991	9990	9990	9989	9988	9987					
3	·9986	9985	9984	9983	9982	9981	9980	9979	9978	9977	0	0	1	1	1
4	·9976	9974	9973	9972	9971	9969	9968	9966	9965	9963	0	0	1	1	1
5	·9962	9960	9959	9957	9956	9954	9952	9951	9949	9947	0	1	1	1	1
6	·9945	9943	9942	9940	9938	9936	9934	9932	9930	9928	0	1	1	1	2
7	·9925	9923	9921	9919	9917	9914	9912	9910	9907	9905	0	1	1	2	2
8	·9903	9900	9898	9895	9893	9890	9888	9885	9882	9880	0	1	1	2	2
9	·9877	9874	9871	9869	9866	9863	9860	9857	9854	9851	0	1	1	2	2
10	·9848	9845	9842	9839	9836	9833	9829	9826	9823	9820	1	1	2	2	3
11	·9816	9813	9810	9806	9803	9799	9796	9792	9789	9785	1	1	2	2	3
12	·9781	9778	9774	9770	9767	9763	9759	9755	9751	9748	1	1	2	3	3
13	·9744	9740	9736	9732	9728	9724	9720	9715	9711	9707	1	1	2	3	3
14	·9703	9699	9694	9690	9686	9681	9677	9673	9668	9664	1	1	2	3	4
15	·9659	9655	9650	9646	9641	9636	9632	9627	9622	9617	1	2	2	3	4
16	·9613	9608	9603	9598	9593	9588	9583	9578	9573	9568	1	2	2	3	4
17	·9563	9558	9553	9548	9542	9537	9532	9527	9521	9516	1	2	3	4	4
18	·9511	9505	9500	9494	9489	9483	9478	9472	9466	9461	1	2	3	4	5
19	·9455	9449	9444	9438	9432	9426	9421	9415	9409	9403	1	2	3	4	5
20	·9397	9391	9385	9379	9373	9367	9361	9354	9348	9342	1	2	3	4	5
21	·9336	9330	9323	9317	9311	9304	9298	9291	9285	9278	1	2	3	4	5
22	·9272	9265	9259	9252	9245	9239	9232	9225	9219	9212	1	2	3	4	6
23	·9205	9198	9191	9184	9178	9171	9164	9157	9150	9143	1	2	3	5	6
24	·9135	9128	9121	9114	9107	9100	9092	9085	9078	9070	1	2	4	5	6
25	·9063	9056	9048	9041	9033	9026	9018	9011	9003	8996	1	3	4	5	6
26	·8988	8980	8973	8965	8957	8949	8942	8934	8926	8918	1	3	4	5	6
27	·8910	8902	8894	8886	8878	8870	8862	8854	8846	8838	1	3	4	5	7
28	·8829	8821	8813	8805	8796	8788	8780	8771	8763	8755	1	3	4	6	7
29	·8746	8738	8729	8721	8712	8704	8695	8686	8678	8669	1	3	4	6	7
30	·8660	8652	8643	8634	8625	8616	8607	8599	8590	8581	1	3	4	6	7
31	·8572	8563	8554	8545	8536	8526	8517	8508	8499	8490	2	3	5	6	8
32	·8480	8471	8462	8453	8443	8434	8425	8415	8406	8396	2	3	5	6	8
33	·8387	8377	8368	8358	8348	8339	8329	8320	8310	8300	2	3	5	6	8
34	·8290	8281	8271	8261	8251	8241	8231	8221	8211	8202	2	3	5	7	8
35	·8192	8181	8171	8161	8151	8141	8131	8121	8111	8100	2	3	5	7	8
36	·8090	8080	8070	8059	8049	8039	8028	8018	8007	7997	2	3	5	7	9
37	·7986	7976	7965	7955	7944	7934	7923	7912	7902	7891	2	4	5	7	9
38	·7880	7869	7859	7848	7837	7826	7815	7804	7793	7782	2	4	5	7	9
39	·7771	7760	7749	7738	7727	7716	7705	7694	7683	7672	2	4	6	7	9
40	·7660	7649	7638	7627	7615	7604	7593	7581	7570	7559	2	4	6	8	9
41	·7547	7536	7524	7513	7501	7490	7478	7466	7455	7443	2	4	6	8	10
42	·7431	7420	7408	7396	7385	7373	7361	7349	7337	7325	2	4	6	8	10
43	·7314	7302	7290	7278	7266	7254	7242	7230	7218	7206	2	4	6	8	10
44	·7193	7181	7169	7157	7145	7133	7120	7108	7096	7083	2	4	6	8	10
	0'	6'	12'	18'	24'	30'	36'	42'	48'	54'	1'	2'	3'	4'	5'

The black type indicates that the integer changes.

	0'	6'	12'	18'	24'	30'	36'	42'	48'	54'	1'	2'	3'	4'	5'
45°	·7071	7059	7046	7034	7022	7009	6997	6984	6972	6959	2	4	6	8	10
46	·6947	6934	6921	6909	6896	6884	6871	6858	6845	6833	2	4	6	8	11
47	·6820	6807	6794	6782	6769	6756	6743	6730	6717	6704	2	4	6	9	11
48	·6691	6678	6665	6652	6639	6626	6613	6600	6587	6574	2	4	7	9	11
49	·6561	6547	6534	6521	6508	6494	6481	6468	6455	6441	2	4	7	9	11
50	·6428	6414	6401	6388	6374	6361	6347	6334	6320	6307	2	4	7	9	11
51	·6293	6280	6266	6252	6239	6225	6211	6198	6184	6170	2	5	7	9	11
52	·6157	6143	6129	6115	6101	6088	6074	6060	6046	6032	2	5	7	9	12
53	·6018	6004	5990	5976	5962	5948	5934	5920	5906	5892	2	5	7	9	12
54	·5878	5864	5850	5835	5821	5807	5793	5779	5764	5750	2	5	7	10	12
55	·5736	5721	5707	5693	5678	5664	5650	5635	5621	5606	2	5	7	10	12
56	·5592	5577	5563	5548	5534	5519	5505	5490	5476	5461	2	5	7	10	12
57	·5446	5432	5417	5402	5388	5373	5358	5344	5329	5314	2	5	7	10	12
58	·5299	5284	5270	5255	5240	5225	5210	5195	5180	5165	2	5	7	10	12
59	·5150	5135	5120	5105	5090	5075	5060	5045	5030	5015	3	5	8	10	13
60	·5000	4985	4970	4955	4939	4924	4909	4894	4879	4863	3	5	8	10	13
61	·4848	4833	4818	4802	4787	4772	4756	4741	4726	4710	3	5	8	10	13
62	·4695	4679	4664	4648	4633	4617	4602	4586	4571	4555	3	5	8	10	13
63	·4540	4524	4509	4493	4478	4462	4446	4431	4415	4399	3	5	8	10	13
64	·4384	4368	4352	4337	4321	4305	4289	4274	4258	4242	3	5	8	11	13
65	·4226	4210	4195	4179	4163	4147	4131	4115	4099	4083	3	5	8	11	13
66	·4067	4051	4035	4019	4003	3987	3971	3955	3939	3923	3	5	8	11	13
67	·3907	3891	3875	3859	3843	3827	3811	3795	3778	3762	3	5	8	11	13
68	·3746	3730	3714	3697	3681	3665	3649	3633	3616	3600	3	5	8	11	14
69	·3584	3567	3551	3535	3518	3502	3486	3469	3453	3437	3	5	8	11	14
70	·3420	3404	3387	3371	3355	3338	3322	3305	3289	3272	3	5	8	11	14
71	·3256	3239	3223	3206	3190	3173	3156	3140	3123	3107	3	6	8	11	14
72	·3090	3074	3057	3040	3024	3007	2990	2974	2957	2940	3	6	8	11	14
73	·2924	2907	2890	2874	2857	2840	2823	2807	2790	2773	3	6	8	11	14
74	·2756	2740	2723	2706	2689	2672	2656	2639	2622	2605	3	6	8	11	14
75	·2588	2571	2554	2538	2521	2504	2487	2470	2453	2436	3	6	8	11	14
76	·2419	2402	2385	2368	2351	2334	2317	2300	2284	2267	3	6	8	11	14
77	·2250	2233	2215	2198	2181	2164	2147	2130	2113	2096	3	6	9	11	14
78	·2079	2062	2045	2028	2011	1994	1977	1959	1942	1925	3	6	9	11	14
79	·1908	1891	1874	1857	1840	1822	1805	1788	1771	1754	3	6	9	11	14
80	·1736	1719	1702	1685	1668	1650	1633	1616	1599	1582	3	6	9	11	14
81	·1564	1547	1530	1513	1495	1478	1461	1444	1426	1409	3	6	9	12	14
82	·1392	1374	1357	1340	1323	1305	1288	1271	1253	1236	3	6	9	12	14
83	·1219	1201	1184	1167	1149	1132	1115	1097	1080	1063	3	6	9	12	14
84	·1045	1028	1011	0993	0976	0958	0941	0924	0906	0889	3	6	9	12	14
85	·0872	0854	0837	0819	0802	0785	0767	0750	0732	0715	3	6	9	12	14
86	·0698	0680	0663	0645	0628	0610	0593	0576	0558	0541	3	6	9	12	15
87	·0523	0506	0488	0471	0454	0436	0419	0401	0384	0366	3	6	9	12	15
88	·0349	0332	0314	0297	0279	0262	0244	0227	0209	0192	3	6	9	12	15
89	·0175	0157	0140	0122	0105	0087	0070	0052	0035	0017	3	6	9	12	15
	0'	6'	12'	18'	24'	30'	36'	42'	48'	54'	1'	2'	3'	4'	5'

	0′	6′	12′	18′	24′	30′	36′	42′	48′	54′	1′	2′	3′	4′	5′
0°	0·0000	0017	0035	0052	0070	0087	0105	0122	0140	0157	3	6	9	12	15
1	0·0175	0192	0209	0227	0244	0262	0279	0297	0314	0332	3	6	9	12	15
2	0·0349	0367	0384	0402	0419	0437	0454	0472	0489	0507	3	6	9	12	15
3	0·0524	0542	0559	0577	0594	0612	0629	0647	0664	0682	3	6	9	12	15
4	0·0699	0717	0734	0752	0769	0787	0805	0822	0840	0857	3	6	9	12	15
5	0·0875	0892	0910	0928	0945	0963	0981	0998	1016	1033	3	6	9	12	15
6	0·1051	1069	1086	1104	1122	1139	1157	1175	1192	1210	3	6	9	12	15
7	0·1228	1246	1263	1281	1299	1317	1334	1352	1370	1388	3	6	9	12	15
8	0·1405	1423	1441	1459	1477	1495	1512	1530	1548	1566	3	6	9	12	15
9	0·1584	1602	1620	1638	1655	1673	1691	1709	1727	1745	3	6	9	12	15
10	0·1763	1781	1799	1817	1835	1853	1871	1890	1908	1926	3	6	9	12	15
11	0·1944	1962	1980	1998	2016	2035	2053	2071	2089	2107	3	6	9	12	15
12	0·2126	2144	2162	2180	2199	2217	2235	2254	2272	2290	3	6	9	12	15
13	0·2309	2327	2345	2364	2382	2401	2419	2438	2456	2475	3	6	9	12	15
14	0·2493	2512	2530	2549	2568	2586	2605	2623	2642	2661	3	6	9	12	16
15	0·2679	2698	2717	2736	2754	2773	2792	2811	2830	2849	3	6	9	13	16
16	0·2867	2886	2905	2924	2943	2962	2981	3000	3019	3038	3	6	9	13	16
17	0·3057	3076	3096	3115	3134	3153	3172	3191	3211	3230	3	6	10	13	16
18	0·3249	3269	3288	3307	3327	3346	3365	3385	3404	3424	3	6	10	13	16
19	0·3443	3463	3482	3502	3522	3541	3561	3581	3600	3620	3	7	10	13	16
20	0·3640	3659	3679	3699	3719	3739	3759	3779	3799	3819	3	7	10	13	17
21	0·3839	3859	3879	3899	3919	3939	3959	3979	4000	4020	3	7	10	13	17
22	0·4040	4061	4081	4101	4122	4142	4163	4183	4204	4224	3	7	10	14	17
23	0·4245	4265	4286	4307	4327	4348	4369	4390	4411	4431	3	7	10	14	17
24	0·4452	4473	4494	4515	4536	4557	4578	4599	4621	4642	4	7	11	14	18
25	0·4663	4684	4706	4727	4748	4770	4791	4813	4834	4856	4	7	11	14	18
26	0·4877	4899	4921	4942	4964	4986	5008	5029	5051	5073	4	7	11	15	18
27	0·5095	5117	5139	5161	5184	5206	5228	5250	5272	5295	4	7	11	15	18
28	0·5317	5340	5362	5384	5407	5430	5452	5475	5498	5520	4	8	11	15	19
29	0·5543	5566	5589	5612	5635	5658	5681	5704	5727	5750	4	8	12	15	19
30	0·5774	5797	5820	5844	5867	5890	5914	5938	5961	5985	4	8	12	16	20
31	0·6009	6032	6056	6080	6104	6128	6152	6176	6200	6224	4	8	12	16	20
32	0·6249	6273	6297	6322	6346	6371	6395	6420	6445	6469	4	8	12	16	20
33	0·6494	6519	6544	6569	6594	6619	6644	6669	6694	6720	4	8	13	17	21
34	0·6745	6771	6796	6822	6847	6873	6899	6924	6950	6976	4	9	13	17	21
35	0·7002	7028	7054	7080	7107	7133	7159	7186	7212	7239	4	9	13	18	22
36	0·7265	7292	7319	7346	7373	7400	7427	7454	7481	7508	5	9	14	18	23
37	0·7536	7563	7590	7618	7646	7673	7701	7729	7757	7785	5	9	14	18	23
38	0·7813	7841	7869	7898	7926	7954	7983	8012	8040	8069	5	9	14	19	24
39	0·8098	8127	8156	8185	8214	8243	8273	8302	8332	8361	5	10	15	20	24
40	0·8391	8421	8451	8481	8511	8541	8571	8601	8632	8662	5	10	15	20	25
41	0·8693	8724	8754	8785	8816	8847	8878	8910	8941	8972	5	10	16	21	26
42	0·9004	9036	9067	9099	9131	9163	9195	9228	9260	9293	5	11	16	21	27
43	0·9325	9358	9391	9424	9457	9490	9523	9556	9590	9623	6	11	17	22	28
44	0·9657	9691	9725	9759	9793	9827	9861	9896	9930	9965	6	11	17	23	28
	0′	6′	12′	18′	24′	30′	36′	42′	48′	54′	1′	2′	3′	4′	5′

	0'	6'	12'	18'	24'	30'	36'	42'	48'	54'	1'	2'	3'	4'	5'
45°	1·0000	0035	0070	0105	0141	0176	0212	0247	0283	0319	6	12	18	24	30
46	1·0355	0392	0428	0464	0501	0538	0575	0612	0649	0686	6	12	18	25	31
47	1·0724	0761	0799	0837	0875	0913	0951	0990	1028	1067	6	13	19	25	32
48	1·1106	1145	1184	1224	1263	1303	1343	1383	1423	1463	7	13	20	26	33
49	1·1504	1544	1585	1626	1667	1708	1750	1792	1833	1875	7	14	21	28	34
50	1·1918	1960	2002	2045	2088	2131	2174	2218	2261	2305	7	14	22	29	36
51	1·2349	2393	2437	2482	2527	2572	2617	2662	2708	2753	8	15	23	30	38
52	1·2799	2846	2892	2938	2985	3032	3079	3127	3175	3222	8	16	24	31	39
53	1·3270	3319	3367	3416	3465	3514	3564	3613	3663	3713	8	16	25	33	41
54	1·3764	3814	3865	3916	3968	4019	4071	4124	4176	4229	9	17	26	34	43
55	1·4281	4335	4388	4442	4496	4550	4605	4659	4715	4770	9	18	27	36	45
56	1·4826	4882	4938	4994	5051	5108	5166	5224	5282	5340	10	19	29	38	48
57	1·5399	5458	5517	5577	5637	5697	5757	5818	5880	5941	10	20	30	40	50
58	1·6003	6066	6128	6191	6255	6319	6383	6447	6512	6577	11	21	32	43	53
59	1·6643	6709	6775	6842	6909	6977	7045	7113	7182	7251	11	23	34	45	56
60	1·7321	7391	7461	7532	7603	7675	7747	7820	7893	7966	12	24	36	48	60
61	1·8040	8115	8190	8265	8341	8418	8495	8572	8650	8728	13	26	38	51	64
62	1·8807	8887	8967	9047	9128	9210	9292	9375	9458	9542	14	27	41	55	68
63	1·9626	9711	9797	9883	9970	**0057**	**0145**	**0233**	**0323**	**0413**	15	29	44	58	73
64	2·0503	0594	0686	0778	0872	0965	1060	1155	1251	1348	16	31	47	63	78
65	2·1445	1543	1642	1742	1842	1943	2045	2148	2251	2355	17	34	51	68	85
66	2·2460	2566	2673	2781	2889	2998	3109	3220	3332	3445	18	37	55	73	91
67	2·3559	3673	3789	3906	4023	4142	4262	4383	4504	4627	20	40	60	79	99
68	2·4751	4876	5002	5129	5257	5386	5517	5649	5782	5916	22	43	65	87	108
69	2·6051	6187	6325	6464	6605	6746	6889	7034	7179	7326	24	47	71	95	119
70	2·7475	7625	7776	7929	8083	8239	8397	8556	8716	8878	26	52	78	104	130
71	2·9042	9208	9375	9544	9714	9887	**0061**	**0237**	**0415**	**0595**	29	58	87	116	144
72	3·0777	0961	1146	1334	1524	1716	1910	2106	2305	2506	32	64	97	129	161
73	3·2709	2914	3122	3332	3544	3759	3977	4197	4420	4646	36	72	108	144	180
74	3·4874	5105	5339	5576	5816	6059	6305	6554	6806	7062	41	81	122	163	203
75	3·7321	7583	7848	8118	8391	8667	8947	9232	9520	9812	46	93	139	186	232
76	4·0108	0408	0713	1022	1335	1653	1976	2303	2635	2972	53	107	160	214	267
77	4·3315	3662	4015	4373	4737	5107	5483	5864	6252	6646	62	124	186	248	310
78	4·7046	7453	7867	8288	8716	9152	9594	**0045**	**0504**	**0970**	73	146	220	293	366
79	5·1446	1929	2422	2924	3435	3955	4486	5026	5578	6140	87	175	263	350	438
80	5·671	5·730	5·789	5·850	5·912	5·976	6·041	6·107	6·174	6·243					
81	6·314	6·386	6·460	6·535	6·612	6·691	6·772	6·855	6·940	7·026					
82	7·115	7·207	7·300	7·396	7·495	7·596	7·700	7·806	7·916	8·028					
83	8·144	8·264	8·386	8·513	8·643	8·777	8·915	9·058	9·205	9·357					
84	9·51	9·68	9·84	10·02	10·20	10·39	10·58	10·78	10·99	11·20	Differences				
85	11·43	11·66	11·91	12·16	12·43	12·71	13·00	13·30	13·62	13·95	untrustworthy				
86	14·30	14·67	15·06	15·46	15·89	16·35	16·83	17·34	17·89	18·46	here				
87	19·08	19·74	20·45	21·20	22·02	22·90	23·86	24·90	26·03	27·27					
88	28·64	30·14	31·82	33·69	35·80	38·19	40·92	44·07	47·74	52·08					
89	57·29	63·66	71·62	81·85	95·49	114·6	143·2	191·0	286·5	573·0					
	0'	6'	12'	18'	24'	30'	36'	42'	48'	54'	1'	2'	3'	4'	5'

The black type indicates that the integer changes.

INDEX

Algebraic multiplication and division, 45, 47
Algebraic sums, 43
Angles, 65
  complementary, 67, 190, 195
  in circles, 199
  supplementary, 67
Annulus, area of, 117
Antilogarithms, 103
Averages, 34

Brackets, 49

Cancelling, 8
Centre-finder, engineer's, 198
Change of subject, 62
Characteristic, 100
Chords, 196, 208
Circle, area of, 114
  circumference, 110
  definitions, 196
Circular pitch, 113
Coefficients, 42
Cone, 13, 16, 163
Congruence, 79
Co-secant, 195
Cosine, 190
Co-tangent, 195
Cube, surface and volume, 149
Cylinder, 13, 16, 153

Decimals, 21
  place value, 21
Degree, 65
Diagonal Scales, 216

Elevation, angle of, 187
Ellipse, 212
Equation, simple, 57
  quadratic, 179
  simultaneous, 127, 136
Equilateral triangle, 73, 95, 96

Factors, 1, 52
Fractions, 3
Frustum of Pyramid, 165, 236

Graphs, 119

Highest common factor, 1

Inverse proportion, 32

Loci, 213
Logarithms, 100
Lowest common multiple, 2

Mantissa, 101
Multiples, 2

Parabola, 176
Parallel lines, 70
Parallelogram, 82
Percentages, 36
Plane figures, 15
Planes, 14
Polygons, 75, 88, 96, 233
Powers, 39, 45
Prime numbers, 1
Prisms, 13, 147, 152
Pyramids, 13, 147, 160, 236
Pythagoras, theorem of, 89, 97

Quadratic equations, 179
Quadrilaterals, 82, 86

Radial projection, 170
Ratio, 30, 170
Rectangle, area of, 17, 82
Roots, 39, 97

Secant, 195, 202
Sector, area of, 116
Segment of circle, 196
Set-squares, 67, 94, 95
Significant figures, 24
Signs, rule of, 46
Similar figures, 170
Similar solids, 174
Sine of an angle, 189
Slope of a graph, 139
Sphere, 13, 16, 167, 236
Square roots graphically, 97, 208

Tables and constants, 231
Tangents, 202
Transportation, 58, 62
Trapezium, 82, 85
Triangle, area of, 19, 83
Trigonometry, 185

Volume, 13, 148

ANSWERS

Exercises 1

1. 21. **2.** 14. **3.** 3. **4.** 3.

5. 15. **6.** 10. **7.** 120. **8.** 105.

9. 14. **10.** 24. **11.** 30. **12.** 19.

13. 5 in.; 49. **14.** 1 ft. 6 in. square.

Exercises 2

1. 96. **2.** 240. **3.** 5400. **4.** 36.

5. 630. **6.** 3360. **7.** 15. **8.** 7.

9. Large wheel 4; small wheel 7. **10.** 12.

Exercises 3

1. $5\frac{5}{8}$, $3\frac{2}{7}$, $1\frac{8}{9}$, $2\frac{2}{5}$, $16\frac{3}{4}$, $6\frac{1}{3}$. **2.** $8\frac{1}{18}$, $10\frac{7}{23}$, $21\frac{17}{19}$, $20\frac{6}{41}$, $4\frac{11}{17}$.

3. $\frac{75}{8}$, $\frac{123}{7}$, $\frac{94}{15}$, $\frac{259}{12}$, $\frac{63}{4}$. **4.** $\frac{111}{11}$, $\frac{251}{13}$, $\frac{180}{7}$, $\frac{130}{9}$, $\frac{187}{10}$.

Exercises 4

1. $\frac{5}{21}$, $\frac{3}{5}$, $5\frac{1}{5}$, $\frac{2}{9}$. **2.** $1\frac{1}{11}$, $\frac{6}{13}$, $\frac{5}{19}$, $2\frac{3}{4}$.

Exercises 5

1. $\frac{1}{9}$, $\frac{4}{27}$, $\frac{11}{12}$, $\frac{80}{81}$; $1\frac{8}{81}$. **2.** 3.

3. (a) $4\frac{3}{4}$, (b) $9\frac{5}{8}$, (c) $11\frac{1}{4}$, (d) 1 ft. $0\frac{1}{4}$ in. **4.** $1\frac{8}{15}$.

5. $\frac{7}{8}$. **6.** $5\frac{13}{48}$.

7. 1. **8.** $11\frac{3}{40}$.

9. $1\frac{41}{63}$. **10.** $11\frac{1}{2}$.

11. $14\frac{5}{24}$. **12.** $\frac{109}{120}$.

13. $17\frac{3}{4}$ yd. **14.** $2\frac{5}{24}$ miles.

15. Copper 192 lb.; tin 16 lb.

Exercises 6

1. $\frac{1}{4}$. **2.** $6\frac{1}{8}$. **3.** $6\frac{1}{8}$. **4.** $\frac{1}{10}$.

5. $1\frac{7}{12}$. **6.** $\frac{3}{28}$. **7.** $\frac{9}{10}$. **8.** $\frac{3}{8}$.

9. 20. **10.** $\frac{3}{4}$.

Exercises 7

1. $1\frac{1}{4}$.   **2.** $1\frac{1}{4}$.   **3.** $\frac{1}{6}$.   **4.** $2\frac{23}{63}$.
5. $3\frac{3}{8}$.   **6.** $\frac{31}{51}$.   **7.** $3\frac{11}{12}$.   **8.** $2\frac{5}{24}$.
9. $2\frac{17}{30}$.   **10.** $\frac{1}{14}$.

Exercises 8

1. 160.   **2.** $\frac{127}{241}$.   **3.** 24.   **4.** $\frac{9}{19}$.
5. $1\frac{16}{47}$.   **6.** $\frac{35}{102}$.   **7.** $1\frac{4}{5}$.   **8.** $\frac{17}{31}$.
9. 42.   **10.** $\frac{2}{25}$.

Exercises 9

1. $\frac{1}{3}$.   **2.** $\frac{1}{8}$.   **3.** $\frac{20}{33}$.   **4.** $\frac{4}{7}$.
5. $\frac{2}{5}$.   **6.** $\frac{3}{13}$.   **7.** $\frac{597}{1673}$.   **8.** $\frac{5}{66}$.
9. £0. 7s. 6d.   **10.** 132 yd.

Exercises 11

1. $55\frac{1}{4}$ sq. ft.
2. (a) 44 ft.; (b) 120 sq. ft.; (c) 352 sq. ft.
3. $48\frac{7}{8}$ sq. ft.   **4.** 324 sq. ft.
5. $216\frac{1}{12}$ sq. ft.; 64 ft. 10 in.
6. $6\frac{15}{32}$ sq. in.   **7.** (a) $218\frac{2}{3}$ sq. ft.; (b) 62 ft. 8 in.
8. 250 sq. in.; 6 ft.   **9.** $326\frac{1}{8}$ sq. ft.; 81 ft. 6 in.

Exercises 12

1. 99 sq. ft.; 24 sq. ft.   **2.** £3. 13s. 4d.
3. $409\frac{19}{48}$ sq. ft.   **4.** 1280 sq. ft.

Exercises 13

1. 299·7313.   **2.** ·39488.   **3.** 72·3265.   **4.** 1·9584.
5. ·28408.   **6.** ·0100648.   **7.** 1·741025.   **8.** 18·68679.
9. ·35232.   **10.** 2·4192.   **11.** 8·616.   **12.** 1214·464.
13. 1·9824.   **14.** 122·1745.   **15.** ·0455.   **16.** ·035.
17. 0·054.   **18.** ·00063.   **19.** ·005.   **20.** ·00006.

Exercises 14

1. 3·6.   **2.** 19.   **3.** 84·89.   **4.** 137.

5. 7·3.   **6.** 22·72.   **7.** 26·5.   **8.** 3·01.

9. 12.   **10.** 30,100.   **11.** ·00301.   **12.** ·826.

13. ·00005.   **14.** 1·8.   **15.** ·002.   **16.** 1112.

17. 0·394   **18.** ·000008.   **19.** ·000008.   **20.** 1,000,000.

Exercises 15

1. ·53333.   **2.** ·77604.   **3.** 0·69583.   **4.** ·32292.

5. ·56875.   **6.** ·14896.   **7.** 0·3875.   **8.** ·02813.

9. ·19375.   **10.** ·92813.   **11.** £0. 8s. $6\frac{3}{4}d$.   **12.** £0. 1s. $11\frac{3}{4}d$.

13. £3. 5s. $11\frac{1}{2}d$.   **14.** 3 tons 17 cwt. 1 qr. 1 lb.

15. 5 cwt. 2 qr. 15 lb.   **16.** 2 ft. 5 in.

17. 682 yd.   **18.** 1185 yd. 1 ft. 1 in.

19. 0·0156.   **20.** ·075.

Revision Exercises

1. 48.   **2.** 720.   **3.** $\frac{3}{8}$.

4. 4 ft.   **5.** £17. 8s. 6d.   **6.** 3 times.

7. £129. 12s. 0d.   **8.** 2, 3, 7, 11, 11, 31.   **9.** 14.

10. 4200 lb.   **11.** 5 pints.   **12.** 956.

13. 3 miles.   **14.** £1. 5s. $11\frac{1}{2}d$.   **15.** £65.

16. $4\frac{3}{4}$.   **17.** 144.   **18.** £0. 12s. $9\frac{3}{4}d$.

19. 213 pieces; $1\frac{1}{8}$ in.   **20.** 5, 3, 2.   **21.** 15 yd.

22. 105 gall.   **23.** 17·5.   **24.** 17.

25. 8.   **26.** £2. 8s. $8\frac{3}{4}d$.   **27.** 3 ft. $10\frac{27}{32}$ in.

28. $46\frac{2}{3}$ lb.   **29.** 2·775.   **30.** 117 yd. 0 ft. 5·4 in.

Exercises 16

1. 90.   **2.** 10.   **3.** 1 lb.

4. 107 pistons. **5.** £0. 17s. 0d.   **6.** £0. 5s. $7\frac{1}{2}d$.

7. $2\frac{1}{2}$ dozen.   **8.** £18. 18s. 3d.   **9.** $\frac{7}{20}$.

10. £0. 2s. 0d. **11.** £44.   **12.** £1050.

13. £415. 6s. 8d.   **14.** 5340 cu. ft.   **15.** £20. 14s. 7d.

16. £0. 6s. $2\frac{1}{4}d$.   **17.** £6. 11s. 3d.   **18.** £2. 5s. 0d.

19. 1026 miles. **20.** 30·25 lb.

Exercises 17

1. 12 men.   **2.** 72 yd.   **3.** 1440 revs.   **4.** 40 m.p.h.

5. 4 ft. 6 in.   **6.** 1500 lb.   **7.** 6 in.   **8.** 1·16 ohm.

9. $\frac{5}{16}$ in.   **10.** 506 sq. ft.

Exercises 18

1. 19.   **2.** 27.   **3.** $4\frac{7}{8}$.

4. £0. 5s. 10d.   **5.** 94.   **6.** £0, 3s. $9\frac{3}{4}d$.

7. 32·8.   **8.** £0. 2s. 8d.   **9.** 13 st. 5 lb.

10. Nil.   **11.** $2\frac{1}{4}d$.   **12.** $\frac{2}{27}$ in.

13. 39.   **14.** 89·24.   **15.** £4. 14s. 11d.

Exercises 19

1. 25; $37\frac{1}{2}$; 70; $41\frac{2}{3}$; $13\frac{1}{3}$; $20\frac{5}{6}$; $21\frac{7}{8}$; $17\frac{7}{9}$.

2. $\frac{4}{25}$; $\frac{1}{3}$; $\frac{9}{20}$; $\frac{1}{40}$; $\frac{3}{4}$; $\frac{1}{20}$; $\frac{7}{25}$; $\frac{29}{180}$.

3. £52. 16s. 7d.   **4.** 308 yd.   **5.** 4 gall.

6. 336 lb.   **7.** $2\frac{1}{2}$ p.c.   **8.** $3\frac{1}{8}$ p.c.

9. £47. 19s. 2d.   **10.** 38.   **11.** $8\frac{8}{9}$ p.c.

12. 171 gall.   **13.** £192. 2s. 3d.   **14.** £700.

15. 21 lb.   **16.** $7\frac{9}{13}$ p.c.   **17.** $28\frac{1}{3}$, $43\frac{1}{3}$, $4\frac{1}{6}$, $18\frac{3}{4}$, $5\frac{5}{12}$

18. $6\frac{1}{4}$ p.c.   **19.** $11\frac{1}{9}$ p.c.   **20.** £1035.

Exercises 20

1. 77.   **2.** 84.   **3.** 71.

4. 34.   **5.** 92.   **6.** 7·483.

7. 9·327.   **8.** 16·941.   **9.** 12·4.

10. 29·01.   **11.** 60·8.   **12.** ·25.

13. ·913.   **14.** ·612.   **15.** ·745.

16. 1·3416.   **17.** 1·512.   **18.** 2·305.

19. 2·717.   **20.** 3·0183.   **21.** 1·773.

Exercises 21

1. 3.   **2.** 800.   **3.** 7.

4. 43.   **5.** 0.   **6.** $t = 1\frac{1}{4}$.

7. $V = 146\cdot4$.   **8.** $T = 7\cdot85$.   **9.** $S = 175\cdot2$.

10. $S = 17\cdot03$.

ANSWERS

Exercises 22

1. $-3x+2y.$ 2. $-2a+3c.$
3. $2x^2-2x-4.$ 4. $12-6x+x^2.$
5. $x^2+4x-2.$ 6. $p^3-p^2+5.$
7. $\frac{7}{4}a^2+\frac{5}{4}a-4.$ 8. $-3p^2+2p-2q+5.$
9. $x^3+2x.$ 10. $\frac{14}{5}ab.$

Exercises 23

1. $15a^3x^2.$ 2. $-6x^3y^3.$
3. $6x^3y^3-2xy^5.$ 4. $x^4+x^2-72.$
5. $a^3+x^3.$ 6. $12a^2-23ab+10b^2.$
7. $x^3-y^3.$ 8. $x^3-3x+2.$
9. $4x^4+6x^3-8x^2-18x-9.$ 10. $\frac{1}{4}a^3-\frac{4}{3}a^2-3a+12.$
11. $b.$ 12. $-5x^2y.$
13. $7y^2z.$ 14. $a^2+4a+5.$
15. $x^2+3x-4.$ 16. $9+9x+3x^2.$
17. $x^3+x^2+2x+4.$ 18. $4x^3-8x^2+9x-9.$
19. $x^2-4.$ 20. $a^2-4ab+4b^2.$

Exercises 24

1. $a-5b.$ 2. $3x-11y+2.$
3. $3a-5b.$ 4. $x.$
5. $-5.$ 6. $19.$
7. $1.$ 8. $8.$
9. $3a-5b.$ 10. $0.$
11. $x-3y-4(x+2y-3).$ 12. $a-(2b-c)+3(2a-3b).$
13. $(z-4)-(3x+y).$ 14. $28-(x+2y).$
15. $24(f+m).$ 16. $2\{5(m-n)-xy\}.$
17. $\frac{3}{4}(m+2-w).$ 18. $£(4x-3).$
19. $(19-3m)$ miles.   20. $2(x+y+s).$

Exercises 25

1. $x(x+2p-4).$ 2. $3p(1+2p-5p^2).$
3. $\frac{x^2}{y^2}\left(1-\frac{2x}{y}+\frac{3x^2}{y^2}\right).$ 4. $(a+b)(c+a).$
5. $(p+q)(r-s).$ 6. $(a-2)(3x+y).$

Exercises 25 (*cont.*)

7. $(x+8)(x+2)$.

8. $(x-5)(x+3)$.

9. $(x+7)(x-6)$.

10. $(x+7)(x-5)$.

11. $3(a-5)(a+4)$.

12. $(x+12a)(x-6a)$.

13. $2(a-2)(a+1)$.

14. $(ab+6)(ab-8)$.

15. $(a-5)^2$.

16. $(2a+3)^2$.

17. $(3p-s)^2$.

18. $(6x-5y)(6x+5y)$.

19. 9996.

20. 999,984.

Exercises 26

1. $(2x+1)(x+2)$.

2. $(3x+1)(x+2)$.

3. $(3x-1)(2x-1)$.

4. $(x+2)(2x-1)$.

5. $(x+12)(17x-3)$.

6. $(3x+2)(2x-5)$.

7. $(3x+5)(2x-7)$.

8. $(4x+7)(2x-3)$.

9. $(x+12)(4x-7)$.

10. $(3x+1)(5x-8)$.

11. $(3x+2)(4x-1)$.

12. $(x-2y+2k)(x-2y-2k)$.

13. $(a+b+c)(a+b-c)$.

14. $(a-b+c)(a-b-c)$.

15. $5x(x+2)$.

16. $(1+a+b)(1-a-b)$.

17. $(2a+x-y)(2a-x+y)$.

18. $(2a-y+x)(2a-y-x)$.

19. $(3x^2+a^2-b^2)(3x^2-a^2+b^2)$.

20. $(2+x-2y)(2-x+2y)$.

Exercises 27

1. $x=8$.

2. $x=4$.

3. $x=3$.

4. $x=2$.

5. $x=1\frac{1}{2}$.

6. $x=10$.

7. $x=12$.

8. $x=-8$.

9. $x=5$.

10. $x=7$.

11. $x=2\frac{1}{2}$.

12. $x=2$.

13. $x=6$.

14. $x=4$.

15. $x=3\frac{1}{3}$.

16. $x=1$.

17. $x=5$.

18. $x=2$.

19. $x=3$.

20. $x=4$.

Exercises 28

1. $x=80$.

2. $x=30$.

3. $x=22$.

4. $x=4\frac{1}{3}$.

5. $x=5$.

6. $x=8$.

7. $x=8$.

8. $x=5$.

9. $x=27$.

10. $x=6$.

11. $x=4$.

12. $x=7$.

13. $x=\frac{1}{7}$.

14. $x=15$.

15. $x=3$.

ANSWERS 255

Exercises 29

1. 23, 24, 25. **2.** 15, 17, 19. **3.** 7. **4.** 105.

5. 8, 10, 12. **6.** 6s. 3d., 3s. 9d. **7.** £19.

8. 25. **9.** 7 gall. **10.** 120. **11.** 25.

12. £3. **13.** 10. **14.** 20, 25.

15. No difference. **16.** 16 ft.

17. 120 florins, 72 half crowns. **18.** 72, 60, 50.

19. 3s. 9d., 5s. 3d. **20.** 450 sq. ft.

Exercises 30

1. $T = 2$. **2.** $V = 72$.

3. $r = 7 \cdot 98$. **4.** $l = \dfrac{gt^2}{4\pi^2}$.

5. $G = 31 \cdot 2$. **6.** $n = 0 \cdot 0375$.

7. $R = \dfrac{PH}{2\pi n\,(P - W)}$. **8.** $T = 44$.

9. $s = 2$. **10.** $M = 2 \cdot 21$.

11. $V = 2\sqrt{\dfrac{2M}{4m + rs}}$; $V = 1$. **12.** $R = \sqrt{\dfrac{2V}{\pi h} - \dfrac{h^2}{3}}$; $R = 1 \cdot 414$.

Exercises 31

2. Sum $= 360°$. **3.** $90°$, $180°$, $90°$, $0°$. **4.** $210°$.

5. (a) $150°$, $120°$, $88°$, $1°$; (b) $69°$, $45°$, $15°$, $9°$.

6. $140°$, $40°$, $140°$. **7.** $6°$ per min.; $\frac{1}{2}°$ per min. **8.** $11\frac{1}{4}°$.

Exercises 32

1. Equal. 1, 3, 5, 7; 2, 4, 6, 8. Supp. 2, 5; 3, 8, etc.

2. $138°$. **4.** $90°$. **5.** $60°$. **6.** $110°$. **9.** $60°$.

Exercises 33

1. $540°$, $720°$, $900°$, $1080°$. **2.** $72°$, $60°$, $51\frac{3}{7}°$, $45°$.

3. $160°$. **4.** $540° - (4a + 9°)$.

5. $105°$. **6.** $40°$, $110°$.

Exercises 37

1. $3\frac{1}{3}$ in. **3.** 19,800 sq. yd. **5.** $10 \cdot 392$ sq. in.

6. $90°$, $62°$, $90°$, $118°$. **7.** $5\frac{29}{32}$ sq. in. **8.** 1875 sq. ft.

9. 244 sq. ft. **10.** 45 sq. ft. **11.** $4 \cdot 44$ in.; $8 \cdot 88$ in.

Exercises 38

1. 13·854 sq. in.
2. 13·86 sq. in.

Exercises 39

1. 5 in.
2. (1) 2·236 in.; (2) 35 cm.; (3) 104 ft.
3. 70·34 ft.
4. 12·5 cm.
5. 12 ft. 10 in. (approx.).
6. (a) $x^2 + 16x$; (b) $\sqrt{2x^2 + 32x + 256}$.
7. $\sqrt{x^2 - 4}$ (side); $\sqrt{2} \times \sqrt{x^2 - 4}$ (diagonal).
8. 260 ft.
9. 5·304 sq. in.
10. 15·2; 14·16; 12·81 ft.
11. (a) 32·34 in.; (b) 54·98 sq. in.
12. $OX = 4$ ft. 10·8 in.; $OY = 3$ ft. 5·24 in.

Exercises 40

1. 5·66 in.
2. 3·96 in.
3. 4·24 in.; 18 sq. in.
4. 5·656 in.
5. 3·464 in.
6. 28·28 ft.
7. 5·464 ft.

Exercises 41

1. 86·6 sq. in.
2. (a) 27·71 sq. in.; (b) 6·93 in.; (c) 4·62 in.
3. £19. 9s. 8·4d.
4. 23·382 sq. in.; 17·3 p.c.
5. 9·9 in.
6. 14·14 in.
7. 6·62 in.; 39·7 sq. in.
8. 0·086 in.

Exercises 42

1. 3·7542, 2·7542, 1·7542, 0·7542.
2. 0·5452, $\bar{1}$·5452, $\bar{2}$·5452, $\bar{3}$·5452.
3. 1·5887, $\bar{2}$·5887, 5·5887.
4. 0·162, 0·4979, $\bar{1}$·8951.
5. 1·4784, $\bar{4}$·7782, 2·1937.

Exercises 43

1. 11·8, 10·12, 13·28.
2. 144·5, 0·1061, 0·3142.
3. 5, 50, 0·5.
4. 49·16, 138·7, 3·142.
5. 2·316, 0·1002, 0·03221.

ANSWERS

Exercises 44

1. 3·525.   **2.** 85·88.   **3.** 9·674.   **4.** 0·167.
5. 0·7161.   **6.** 0·1064.

Exercises 45

1. 4·894.   **2.** 2·998.   **3.** 0·1966.
4. 12·31.   **5.** 0·08418.   **6.** 0·553.

Exercises 46

1. 18·97.   **2.** 5475.   **3.** 15·55.   **4.** 1·603.
5. 1·51.   **6.** ·01893.   **7.** 134·6.   **8.** 7·532.
9. 2·809.   **10.** 4·638.   **11.** 2·057.   **12.** 54·96.
13. 191·5.   **14.** 6·992.   **15.** 41·26.   **16.** 19·39.
17. 1495.   **18.** 0·128.   **19.** 11·99.   **20.** 1·204.

Exercises 47

1. 5524·8 ft.   **2.** 22 ft.; 120 revs.
3. 22 ft. 4 in.   **4.** 110 ft.
5. (a) $1\frac{1}{4}$; (b) $1\frac{9}{16}$; (c) $2\frac{1}{4}$; (d) $2\frac{7}{8}$ in.   **6.** 702 ft. $10\frac{1}{2}$ in. (approx.).

Exercises 48

1. 2·387 in.   **2.** 11·932 in.   **3.** $\frac{7}{8}$ in.
4. $1\frac{3}{8}$ in.   **5.** 8·59 in.   **6.** 36 teeth.

Exercises 49

1. 22·29 sq. ft.; 2·48 sq. yd.   **2.** 14·96 lb. per sq. in.
3. 127·43 sq. ft.   **4.** (a) 11·28 in.; (b) 42 ft.
5. 61,600 sq. yd.   **6.** 3528 sq. in.
7. 9·142 sq. in.   **8.** £7. 5s. 6d.

Exercises 50

1. 14·46 ft.   **2.** (a) 3·49 in.; (b) 8·73 sq. in.   **3.** 23°.
4. 1·086; 3·976; 1·086; 3·169; 2·714; 2·65; 3·961.
5. £2. 17s. 10d.   **6.** 201·1 sq. in.
7. £25. 2s. 8d.   **8.** (a) 34,048 sq. yd.; (b) £408. 11s. 5d.

Exercises 52

1. $25\frac{1}{2}$ lb.; $10\frac{2}{3}$ in.   **2.** $4\frac{3}{4}$ hours; 13 miles.

3. £11·05.   **4.** 5s. 3d.

5. $5\frac{1}{4}$ miles from house.   **6.** 3·65 ft.

7. Their distance apart.   **8.** 23·25 tons.

9. 6·7, 7·7, 8·4.

Exercises 53

1. $x=4$, $y=4$.   **2.** $x=4$, $y=1$.   **3.** $a=2$, $b=2\frac{1}{2}$.

4. $x=6$, $y=4$.   **5.** $a=11$, $b=9$.   **6.** $r=5$, $p=3$.

7. $x=1$, $y=2$.   **8.** $x=\frac{1}{2}$, $y=2$.   **9.** $x=4$, $y=2$.

10. $x=5$, $y=24$.   **11.** $x=4$, $y=7$.   **12.** $x=4$, $y=3$.

13. $x=5$, $y=8$.   **14.** $x=1\frac{3}{4}$, $y=1\frac{1}{2}$.   **15.** $x=4$, $y=6$.

16. $x=4\frac{1}{2}$, $y=3$.   **17.** $x=2\frac{1}{2}$, $y=1\frac{1}{2}$.   **18.** $x=5$, $y=6$.

19. $x=2$, $y=3$.   **20.** $x=5\frac{1}{2}$, $y=2\frac{1}{2}$.

Exercises 54

1. 8, 6.   **2.** 40, 72.   **3.** 24, 3.   **4.** 7s. 6d

5. $a=0·38$; $b=0·5$; $E=15·7$.   **6.** £7. 12s. 6d.; £4. 19s. 6d

7. 55 at 2s. 6d.; 195 at 1s. 6d.   **8.** 5 days.

9. Rail 12 miles; road 4 miles.   **10.** $Y=2X+3$.

11. $F=1\frac{1}{2}d+\frac{1}{8}$; $5\frac{3}{8}$ in.

12. Boy £1. 12s. 6d.; Man £4. 10s.

13. 33 sixpences, 48 florins.   **14.** 57.

15. 240 miles.

Exercises 55

5. (4·8, 4·5).   **6.** (0, −1).   **7.** $(4\frac{1}{2}, 2\frac{1}{2})$.

Exercises 56

4. 2, 3.   **5.** 3, 2.

Exercises 57

1. $x=2$, $y=3$.   **2.** $x=2$, $y=3$.   **3.** $x=3$, $y=2$.

4. $x=-1$, $y=3$.   **5.** $x=2$, $y=-1$.   **6.** $x=2$, $y=4$.

7. $x=-3$, $y=-2$.   **8.** $x=2\frac{1}{2}$, $y=3$.   **9.** $x=6$, $y=-2$.

10. $x=12$, $y=8$.   **11.** 8, 4.   **12.** 10, 8.

13. 9, 6.   **14.** 12, 9 ft.   **15.** 5, 6.

Exercises 58

1. (i) $y = x - 1$; (ii) $2y + x = 3$; (iii) $x + 2y + 4 = 0$;
   (iv) $y = 2x + 2$; (v) $2y + 3x = 6$; (vi) $2y + 3x = 4$;
   (vii) $5y + 2x = 12$; (viii) $2y = 4x - 5$.

2. $16y = 9x - 2$. **3.** $1 \cdot 05P = 3 \cdot 45 + 0 \cdot 13W$.

4. $y = 1 \cdot 5x - 1$. **5.** $P = 0 \cdot 9 + 0 \cdot 49W$; $4 \cdot 8$ lb.

6. $L = 9 = 2P$. **7.** $y = \frac{3}{2}x - 3$.

8. $y = 6 - 0 \cdot 6x$. **9.** $3 \cdot 142$; $C = 3 \cdot 142D$.

10. $F = 1\frac{1}{2}D + \frac{1}{8}$. **11.** $F = 1 \cdot 8C + 32$.

12. $D = 1 \cdot 2d + 0 \cdot 02$.

Exercises 59

1. (a) 1728 cu. in.; (b) 1 cu. ft. **2.** 66 cu. ft.

3. 55·25 c.cm. **4.** £90; £60.

5. $\frac{15}{16}$ in. **6.** 711 lb.

7. 3·4 lb. **8.** 9·6 in.

Exercises 60

1. 1350 sq. in. **2.** (a) 660 sq. ft.; (b) £1. 13s. 0d.

3. £23. 10s. 0d. **4.** 45 lb.

5. £8. 4s. 7d. **6.** 12·375 cu. in.

Exercises 61

1. 10,400 cu. ft. **2.** $\left(\dfrac{a+b}{2}\right) \times 9xy$ cu. ft.; 16,200 cu. ft.

3. 74·39 lb. **4.** 369·25 cu. yd.

Exercises 62

1. 254·5 sq. in.; 25·45 lb. **2.** 245·4 cu. in.

3. $\frac{3}{14}$. **4.** $5\frac{1}{11}$ in.

5. 0·007 lb. **6.** 4·813 cu. in.; 67·19 cu. in.

7. 1·813 cu. in. **8.** $7\frac{7}{11}$ in.

9. 8·92 lb. **10.** 66 ft. 4 in.

11. 5·093 tons per square inch. **12.** 3367·4 lb.

13. 13,069 lb. **14.** (a) 954 lb.; (b) $3\frac{3}{16}$ in.

Exercises 63

1. (a) 955 cu. in.; (b) 1·704 cu. ft.
2. 0·7854 $(D^2 - d^2)$ t.   3. 395·9 cu. in.; 102·9 lb.
4. 4 lb.   5. 35·71 lb.   6. 160·2 lb.

Exercises 64

1. 8 sq. ft. 55 sq. in. (approx.).   2. 77 sq. ft.
3. $314\frac{2}{7}$ sq. ft.   4. 76·4 in.
5. 1·5708 cu. in.; 5·0944 sq. in.

Exercises 65

1. 17·5 lb.   2. 27·712 cu. in.
3. 27·712 cu. in.   4. 1187·3 lb.

Exercises 66

1. (a) 0·74 cu. in.; (b) 792·2 cu. in.; (c) 20·95 cu. in.
2. 3·77 lb.   3. 6·32 in.; 39·74 sq. in.
4. £67. 17s. 9d.   5. 1072·3 cu. ft.
6. 5236 tons.   7. 16·56 in.
8. 2413 cu. in.

Exercises 67

1. 35·7 cu. in.   2. 76·56 lb.   3. 9242 lb.

Exercises 68

1. 1696 lb.   2. 29·17 cu. ft.
3. 8·995 in.; 381·1 cu. in.   4. 18·7 in.
5. 111·3 cu. in.   6. 5·678 in.; 405·2 sq. in.
7. 0·5236.

Exercises 69

2. 1·732 sq. in.   3. 10·392 sq. in.
4. 117·2 ft.   5. 4.
6. $\frac{1}{25}$.   7. 0·09126 sq. in.

ANSWERS 261

Exercises 70

1. 5·228 in.
2. 3·351 in.
3. 189 lb.
4. 11·53 × 8·784 × 6·039 in.

Exercises 71

3. (3, 9), (2, 4).
5. (2, 4), $(-\frac{1}{2}, \frac{1}{4})$.

Exercises 72

1. 6, −2.
2. 4, −6.
3. 3, −7.
4. 9, −3.
5. 4, 5.
6. −3, −7.
7. 8, −5.
8. 5, −1½.
9. 6, ⅔.
10. ⅕, −2½.
11. 4, 5.
12. 8, −1.
13. 3, −6.
14. −4, −7.
15. 1, 5.
16. 1, 8.
17. 5, 6.
18. 7, −⅓.
19. 2, −1½.
20. 8, −1.
21. ⅘, −2.
22. 4, −¾.
23. 8, −⅔.
24. 14, −3½.
25. ¾, −3½.
26. 10, −¼.
27. ⅔, −24.
28. ±4.
29. 3, −5.
30. 6, 0.
31. ⅕, −1.
32. 7, −⅓.

Exercises 73

1. 5, 12.
2. 7 or 14.
3. 13.
4. 3, 4.
5. 6½, 3½.
6. £8. 5s. 0d.
7. 3 m.p.h.
8. 5, 12.
9. 1·5, 2.
10. 6 yd.
11. 15s.
12. 5 in.
13. 34; 13½ yr.
14. 5 yd.
15. 11, 60.
16. 6 tons.
17. 8, 9.
18. 54 yd.; 1s. 8d.
19. 12, 3.
20. 7 ft.

Exercises 74

1. 3, −2.
2. 4, −3.
3. −1, −2.
4. 2, 8.
5. 1, 4.
6. 3, −5.
7. 1, −½.
8. 1½, −1.
9. 4, −1½.
10. 1, 6.

Exercises 75

1. 0·36, 0·47, 0·58, 0·7, 0·84, 1·0, 1·2, 1·43, 1·73.
3. 21° 48′.
4. 63° 30′.
5. 6·5 ft.
6. 35°.
7. 58°.
8. 9·47 ft.
9. 0·23.
10. $BC = 28·56$ in.; Area = 285·6 sq. in.
11. 28·7 ft.
12. 8·66 ft.
13. 34·3 ft.
14. 8° 7′.
15. 22°.
16. 35°.
17. 21° 48′.
18. 26° 36′.
19. 2 ft. 2 in.; 67° 23′.
20. $DO = 10·46$; $BO = 36·5$.

Exercises 76

1. 53° nearly.

2. 92 ft.

3. 1, 0, ∞.

4. (a) 39 ft.; (b) 67° 23′.

5. 17° 28′.

6. 11·8 ft.

7. 368·5 ft.

8. 36° 53′.

9. 71·7 ft.

10. 0·46 in.

11. 134° 46′.

12. 4·62 cm.

13. 48 ft.

14. $AB = 113\cdot3$; $BC = 52\cdot8$.

15. 12·5 sq. in.

16. $CD = 5\cdot28$; $AD = 12\cdot63$.

17. 30 yd.

18. Tree 34·6 ft.; River 20 ft.

19. 21 ft.

20. 7 ft. 5 in.

Exercises 77

2. 1·12 in.; 0·829 in.

3. 1·98 sq. in.

4. 7·07 ft.

5. 51 in.

Exercises 78

2. 55°, 97½°, 125°, 82½°.

4. 80°, 130°, 100°, 50°; 1·32 in.

5. 29·9 ft.

Exercises 79

2. 120°.

3. 4·43, 4·1, 3·64 in.; 7·014 sq. in.

Exercises 80

1. (a) 2; (b) 3; (c) 1.

2. 2 in.

3. 55°.

5. 2·5, 4·5, 6·5 cm.

Exercises 81

5. $PQA = 55°$; $APQ = 35°$; $QAP = 90°$; $AP = 2\cdot78$ in.; $PQ = 3\cdot4$ in.

Exercises 82

2. 2·449 in.

3. 2·034 in.

5. 8 in.; 12·8 in.

Exercises 83

1. 16·76 in.

2. 6·77 ft.

3. 4·583 cm.

4. 1 : 2.

ANSWERS

TESTS

A

1. 840.
2. £5. 5s. 7$\frac{1}{2}$d.
3. $x = \frac{41}{35}$.
4. 27.
5. 285 sq. ft.; £10. 8s. 6d.
6. (i) 60°, 120°, 120°; (ii) 7·794.

B

1. 6$\frac{3}{7}$.
2. (a) 2·43, 2·87 in.; (b) 2$\frac{11}{32}$, 2$\frac{21}{32}$ in.
3. $x = 5$.
4. 3$\frac{2}{5}$.
5. 88$\frac{1}{4}$ sq. ft.
6. 2·57 in.; 36°, 54°.

C

1. 17 half-crowns, 23 half-sovereigns.
2. $x = 3\frac{1}{2}$, $y = 2\frac{1}{4}$.
3. 115 secs.
4. (i) 0·2208 sq. in.; (ii) 0·3068 sq. in.; (iii) 0·1207 sq. in.
5. 28.
6. 2 in.; 8·5 sq. in.

D

1. £3. 2s. 10d.
2. 17$\frac{1}{2}$ miles; 7 m.p.h.
3. 55461·8 cu. ft.
4. $x = 4$, $y = -5$.
5. £313. 4s. 0d.
6. 2.

E

1. £2. 2s. 9d.
2. (a) $(7x + 3)(7 - 3x)$; (b) $(2 + x)(2 - x)(9 + 4x^2)$.
3. $x = 5$, $y = -1$.
4. $\frac{29}{91}$.
5. 27 ft.
6. $\frac{7}{24}$.

F

1. 4 ft. 6 in.
2. $x = 3$, $y = -2$.
3. 546 lb.
4. 6s. 9d.; 5s.
5. (a) $(a + 3b - c)(a - b + 3c)$; (b) $(x - 2y)(2x + 3y)$.
6. 6·32 in.

G

1. $\dfrac{ax+by}{a+b}$ shillings per lb.

2. (a) $(2x-3)(x+4)$; (b) $(x+y+z)(x+y-z)$.

3. $x=2$, $y=-3$.   4. 3300 cu. in.

5. 99·97.   6. 16 in.

H

1. 21 p.c.   2. $r=\dfrac{E}{C}-R$; 0·45.

3. (i) $x=2$, $y=1$; (ii) 1·24, $-0·52$.   4. 1·558.

5. $\tan A=0·75$.   6. 7 ft. 6 in.; 10 ft.

I

1. £5.   2. $T=2\pi\sqrt{\dfrac{I}{MH}}$.

3. 1·78, $-11·78$.   4. 2·213.

5. 240·15 sq. ft.; 8·4 p.c.

J

1. £11. 5s. 2d.

2. (a) $x=\dfrac{y-c}{m}$; (b) $x=\dfrac{2-3y}{y+3}$; (c) $x=3y^2+2y+1$.

3. 0·03682.   4. $14\frac{4}{7}$ in.

5. $y=x+1$.   6. 63 ft.

K

1. £4. 16s. 0d.   2. $x=4·48$.

3. 1·367.   4. 9·307 p.c.

5. $L=\frac{3}{5}B+\dfrac{188T}{B^2}$.

L

1. $15\frac{5}{13}$ p.c.   2. (a) 1·145; (b) 97·07.

3. 12 in.; 3 in.   4. 1 ft. 4 in.; 36° 53′.

5. £5. 6s. 3d.   6. 4·19.

M

1. 2469.

2. 60 lb. nearly.

3. $x = 3$ or $-1\frac{1}{3}$.

4. $x = 2$ and $x = -2$; (a) -9, (b) -15.

5. $S\sqrt{\dfrac{3fS}{CP}}$.

6. 2s. 6d.

N

1. 1663 lb.

2. (a) 4·41, 1·59; (b) $x = 5\frac{1}{6}$, $y = \frac{2}{5}$.

3. $P = Q\dfrac{(a+2b)}{(2a+b)}$; ·05.

4. 1·831. **5.** £6. 13s. 4d.

6. 62·738 ft.

O

1. 1·2375 ft. **2.** 30·14. **3.** 13 in.

4. 1. **5.** £6. 12s. 0d. **6.** 419·2 ft.

P

1. (a) $x = \frac{5}{6}$, -1; (b) $x = \frac{2}{7}$, $y - \frac{2}{3}$.

2. $P = 0·16W + 2·8$. **3.** (a) 4·53 in.; (b) 1·504 in.

4. (a) 24 ft. 7 in.; (b) 20 ft. 7 in.

5. 2469. **6.** From OA 5 in. From OB 12 in.

Q

1. 1·8 ft. nearly; $1\frac{5}{22}$ m.p.h. **3.** $x = -4$, $y = 3$.

4. $M = 2213$, $H = 0·2361$. **5.** 240. **6.** 3·58.

R

1. 1 : 2. **2.** $50\frac{1}{2}$ gall.

3. (a) $x = 2$; (b) $x = 7$ or $\frac{2}{3}$. **4.** $x = \frac{1}{2}$.

5. A, 3s. 8d.; B, 3s. 6d. **6.** 122°.

S

1. $x = 1 \cdot 22$ and $3 \cdot 28$.

2. $r = \dfrac{144}{D} - 134$; $r = 46$.

3. $44° \, 12'$.

4. $16 \cdot 87$ in.

5. $1 \cdot 082$.

6. £299; $19 \cdot 6$ p.c.

T

1. 57.

2. $11 \cdot 8$ gall.

3. £480 at 4 p.c.; £420 at $4\frac{1}{2}$ p.c.

4. $7 \cdot 9932$.

5. $n = \sqrt{\dfrac{NR - 1}{r}}$; $n = 2\frac{2}{5}$.

6. $y = 0 \cdot 75x + 3$.

U

1. $V = \sqrt{\dfrac{grF}{W}}$; $V = 8$.

2. Top Row: (*a*) $10 \cdot 39$; (*b*) $15 \cdot 365$; (*c*) $6 \cdot 93$, 12, $6 \cdot 93$ (all inches). Bottom Row: (*a*) $8 \cdot 49$ in.; (*b*) $4 \cdot 33$ in.; (*c*) $10 \cdot 464$ ft.

3. $5 \cdot 873$.

4. AB, $y = x$; BC, $5y = x + 72$; CA, $y = 30 - 5x$.

5. 1, 2, 5, -2.

6. 23 ft.

V

1. 266 ft. 8 in.

2. $15 \cdot 71$ cu. in.

3. (i) $(r + 4)(r - 3)$; (ii) $(D + 16)(D - 16)$.

4. $CAB = 20°$; $ABT = 70°$; $BTO = 50°$.

5. (*a*) $\frac{4}{5}$, $\frac{3}{4}$; (*b*) $12 \cdot 2$ cu. ft.

6. (*a*) $4 \cdot 12$ lb.; (*b*) $119 \cdot 5$ lb.

Printed in the United States
By Bookmasters